ARTIFICIAL WORLDS

Computers, Complexity,
and the
Riddle of Life

Other Books by Richard Morris

Light

The End of the World

The Fate of the Universe

Evolution and Human Nature

Dismantling the Universe

Time's Arrows

The Nature of Reality

Assyrians (poetry)

The Edges of Science

Cosmic Questions

Achilles in the Quantum Universe

ARTIFICIAL WORLDS

Computers, Complexity,
and the
Riddle of Life

RICHARD MORRIS

PLENUM TRADE • NEW YORK AND LONDON

ISBN 0-306-46002-5

© 1999 Richard Morris
Plenum Trade is a Division of Plenum Publishing Corporation
233 Spring Street, New York, N.Y. 10013

10 9 8 7 6 5 4 3 2 1

A C.I.P. record for this book is available from the Library of Congress

Printed in the United States of America

CONTENTS

PREFACE

W hat exactly *is* life? A few decades ago, it didn't appear that this was so difficult a question to answer. A living organism was something that absorbed nutrients from its environment, moved, excreted, reproduced, and so on. Although a definition along these lines is still found in many introductory biology texts, it is not very accurate or useful. As I will show in Chapter 1, all of the common definitions of life are inadequate. None of them tells us what a living organism really is. This should not be interpreted as a sign that the science of biology has somehow gotten itself into a state of confusion. A simple analogy should make this clear.

Suppose we ask the question, "What is an atom?" A century ago, this was an easy question. Atoms were conceived to be the indivisible particles of which all matter was composed. Today, of course, we know that matters are not so simple. Atoms have proved not to be indivisible; they can be split into parts. This is something that happens all the time—and I'm not thinking exclusively of nuclear fission. For example, the electrons that produce the picture on your television or computer monitor (they produce light when they strike the screen) were once parts of atoms. The protons that physicists cause to move at high velocities in particle accelerators were once components of atoms. Curiously, there are no atoms in the core of our sun. Temperatures there are far too high to allow electrons to be bound to atomic nuclei.

It is no longer so easy to say precisely what an atom *is*. Our knowledge of atoms is so much greater than it was a hundred years ago that simple definitions no longer work. In order to adequately describe an atom, we must know how the atom's electrons interact with one another and with the atomic nucleus. We must know how an atom absorbs and emits energy. We must understand what causes atoms to bind together into molecules, and so on, and so on. Scientists understand very well how all these processes work. It may be difficult to define the word "atom," but atomic physics no longer stands at the forefront of science. All that remains to be done in the field is the working out of some details.

The situation with respect to the nature of life is similar. If we are to understand living organisms fully, we must know how life was originally created and how it adapts itself to its environment. We must understand the mechanisms that cause evolution to take place. Note that I speak of "mechanisms" in the plural. At the moment there are a number of controversies going on in the field of evolutionary biology. Some scientists are challenging the orthodox notion that natural selection is the only important factor. Creationists sometimes try to make use of these controversies to suggest that evolution itself is being questioned. But nothing could be farther from the truth. All reputable biologists agree that life evolved, and all agree that natural selection (which is sometimes described by the misleading phrase "survival of the fittest") is the most important driving force behind it. However, some suggest that there might be other operative factors as well.

These controversies were already developing when certain scientists began to wonder if a nontraditional approach might shed some light on the workings of evolution. Those who had worked in the field known as "the sciences of complexity" had long been aware that complex systems exhibit emergent properties. That is, the systems behave in ways that cannot be explained by studying their components. For example, a knowledge of the behavior of air molecules is not sufficient to allow one to predict the behavior of hurricanes, and an understanding of the behavior of hydrogen and helium atoms—which make up the bulk of the observable matter in the universe—is not enough if one wants to describe the formation of galaxies. For that matter, scientists cannot explain why a little whirlpool forms when the plug is pulled in a bathtub if they rely only on their knowledge of the behavior of water molecules.

If there is one thing that is obvious about the phenomenon of life, it is that living organisms are complex. Even the simplest bacteria have hundreds of genes, which cause the formation of protein enzymes, which interact in ways too numerous to count. It is not possible to explain the behavior of a bacterium in terms of its component parts. They relate with one another in ways that are much too complicated. Even the simplest creatures are bundles of emergent properties.

Complexity scientists don't spend much time trying to understand how a bacterium works. They happily leave that to the microbiologists. What does concern them is, for example, the question of how life originated in the first place. There are a number of different competing theories about the origin of life. According to some, life began with RNA. The proponents of other theories think that it is more reasonable to postulate that proteins came first. Others think that both RNA and proteins were involved. And there are yet other theories about the matter. But none of these theories seems completely adequate. As a result, research about the origin of life has spawned as many controversies as evolutionary biology.

Complexity scientists approach the problem in a different manner. Some of them wonder if trying to find the particular organic chemicals that were involved in the origin of life might not be the wrong approach. Perhaps, they suggest, life is an emergent property that arises spontaneously when a chemical system attains a certain degree of complexity. As we will see, this is more than just theory. Complex chemicals that exhibit some of the characteristics of life are currently being studied in the laboratory.

If living organisms had come into existence and had never evolved, then life would not be very interesting. Life—at least life on earth—is endowed with the ability to constantly transform itself into something new. Obviously it is impossible to travel back in time to see how this happened. Evolutionary biologists must depend upon the traces left in the fossil record. As a result there are many questions that remain unanswered, including the one that is so often discussed today: Is natural selection all there is to evolution?

If, for example, we discovered some form of extraterrestrial life on one of the other planets of our solar system, or on one of their satellites, certain questions about evolution would be answered. We would be in a much better position to tell what features of terrestrial evolution can be attributed to chance and which were inevitable. Discovering extraterrestrial life might also shed light on a number of other unsolved puzzles. For example, shortly after the first multi-cellular organisms appeared a little over a half-billion years ago, nature engaged in a series of evolutionary experiments which produced organisms that had an amazingly large variety of different forms. Some of these organisms are clearly the ancestors of those that flourish today. But some of the others look quite bizarre to modern eyes (paleontologist Stephen Jay Gould has called them "weird wonders"). No one has ever adequately explained why this "Cambrian explosion" (the Cambrian era is the geological period during which the creatures flourished) happened. Nature has not been so creative at other times. Naturally, biologists have suggested a number of possible explanations, but there is as yet no agreement as to which is most likely to be correct.

If we discovered that some form of extraterrestrial life had also experienced a Cambrian explosion—or, alternatively, that it had not—then it is likely that our understanding of this period of the evolution of life on earth would be enhanced. Sometimes we understand ourselves better if we have a chance to see what others have done, or are doing.

So far, no extraterrestrial life has been discovered. But it does not follow that alien life forms cannot be studied. Complexity scientists have created electronic computer-virus-like life forms that reproduce, mutate, evolve, and die. Since evolution takes place at a much more rapid pace within a computer, where new species can evolve in as little as an hour, it is now possible to study evolution as it happens. Among the phenomena that scientists hope to observe are the evolution of multicellular life forms, an electronic analogue of the Cambrian explosion, and possibly even the evolution of electronic intelligence. Though preliminary research in this area began around 1990, it is only now beginning to proceed at a rapid pace. I, for one, suspect that a number of astonishing new discoveries will be made during the first decade of the twenty-first century.

It is not by chance that evolution is being studied in computers. More often than not, research in the sciences of complexity involves the creation of computer simulations. Some phenomena are simply too complex to understand if we attempt to break them into their component parts. This reductive approach works very well when scientists want to determine the structure of the atomic nucleus. But it doesn't work when one is trying to understand the chemical and biological interactions that go on inside a living organism. However, in a computer simulation, one can put a lot of simple elements together and see how they interact. More often than not, a number of surprising emergent properties will be seen.

I don't want to create the impression that scientists who do research in the sciences of complexity are concerned solely with the phenomenon of life. They are not. The idea that complex systems exhibit emergent properties can be applied in many different fields, ranging from physics to economics. This is why the word "sciences" is used in the plural. In my opinion, the most promising research, and that which has achieved the most striking results, is research that deals with the phenomenon of life.

The sciences of complexity have their critics. Orthodox biologists, in particular, have been especially critical of work in the field. Their attitude most likely stems in part from scientific conservatism. No one who has done research on a topic for years looks kindly upon scientific upstarts who tell them what the truth really is. However, some of their criticisms make telling points, and it would be misleading to write a book of this nature without mentioning them. Every scientific field is full of false starts and wrong turns, of theories that turn out to be inadequate, and of experimental evidence that proves to be not as compelling as it first seemed. The biologists have undoubtedly put their finger on some of these. However, in spite of this, the science of complexity is a field that is full of promise, and experiencing the criticism that is directed at all new scientific endeavors is only likely to make the scientists who work in the field proceed in a more rigorous manner. If some theories turn out to be wrong, that is no great disaster. The history of physics, to cite just one example, is full of failed theories.

The criticism has been anything but unanimous. Some well-known biologists have welcomed the new approach that is being developed. Others have proposed ideas which are very similar to those which the complexity scientists have developed. For example, paleontologists Niles Eldredge and Stephen Jay Gould have suggested there is more to evolution than the action of natural selection on individual organisms. Evolution, they say, also involves phenomena that take place at higher levels of complexity. Neither Eldredge nor Gould is a complexity scientist. Gould has stated quite flatly he lacks the mathematical background to understand some of the theorizing that goes on in the field. Yet he and Eldredge have expressed ideas similar to those encountered in complexity science.

This is supposed to be a book about work in the sciences of complexity. The reader may wonder, therefore, why there is so much in this book about biology, and evolutionary biology in particular. The reason is simple. Biologists have been studying living organisms since the time of Aristotle. They began to entertain ideas about evolution even before the publication of Charles Darwin's *On the Origin of Species*. It is they who discovered the problems that need to be solved. Their controversies provided inspiration for some of the work being done in the sciences of complexity. Writing about complexity without making any mention of the things happening in the science of biology, and evolutionary biology in particular, would lead only to the creation of an inaccurate picture.

In any case, this is not intended to be a book that deals strictly with the complex simulations that are being run on computers. It is a book about life.

1

WHAT IS LIFE?

W hat is life? According to the German Nobel Prize winning biochemist Manfred Eigen, this is not only a very difficult question, it may not even be the right question. "Things we denote as 'living,' " he says, "have too heterogenous characteristics and capabilities for a common definition to give even an inkling of the variety contained within this term."

Eigen may have a point. There has been more scientific activity devoted to the study of life than to any other subject. Life has been studied by physiologists, taxonomists, biochemists, evolutionary biologists, molecular biologists, zoologists, botanists, paleontologists, embryologists, population geneticists, anatomists, and ecologists. And yet scientists do not agree on a definition of life. They don't always agree on the question of what is or is not alive. For example, most would deny that viruses are really "alive," pointing to the fact that a virus consists of nothing but some DNA or RNA surrounded by a protein coat. A virus, they point out, has no metabolism; it makes use of the metabolism of the cells it invades. But there are others who would consider a virus to be an organism "on the edge of life." There are even some scientists who consider computer viruses to be "alive."

For a long time, biologists thought they could define life as anything capable of moving, growing, reproducing, taking nutrients from its environment, excreting, and responding to its environment. Although this might be a good way to explain life to high school students, it really isn't a very useful definition. Some of these proper-

1

ties are seen in things that no one would call "alive," and there are living organisms which lack some of them. A mayfly, for example, doesn't grow or eat; it does this only during its larval stage. Adult mayflies live just long enough to mate and reproduce. And a toaster is certainly more responsive to its environment than, say, a cactus. The toaster can tell when the toast is done and pop it up. Something similar can be said about any device equipped with a remote control. Depending upon the signals that it receives, it will turn itself on, turn itself off, increase or decrease volume, and so on.

Another definition of life sees a living organism as something which exchanges some of its constituents with its environment (by eating or breathing or excreting, for example), without changing any of its general properties. Again, there are exceptions. A seed or a spore can remain dormant for long periods of time. In comparison, a commercial airliner exchanges materials with its environment. It produces energy by "metabolizing" air and fuel, excreting waste gases in the process. Must we then conclude that an airliner is more "alive" than a seed?

Naturally it is possible to invent more technical definitions. For example, one could say living organisms are entities containing information encoded in nucleic acid molecules (DNA and RNA) and controlling chemical reactions by means of enzymes (proteins that act on DNA, RNA, or other proteins). But of course this would exclude any forms of extraterrestrial life that happened to be organized differently. If life exists elsewhere, DNA and RNA might very well be unique to Earth. (See chapters 2 and 3 for more about the structure and roles of DNA and RNA.) Alternatively, one could define life as something that evolves by natural selection. But it has been shown that computer programs can be made to evolve by natural selection. Are they then alive? Some people think they are.

For that matter, natural selection seems to be an operative principle in technological evolution. For example, we have all seen pictures of the early "penny-farthing" bicycle (named after the smallest and largest English copper coins then in circulation). It had a big wheel in front and a small one in back. This was only one of a number of designs that were developed after the bicycle was invented in 1818. Many of the others would look just as bizarre to the modern eye. In

1874 the forerunner of the modern bicycle was invented. It had two wheels of equal size, and the rear wheel was driven by a chain. By 1879, this type had captured the largest share of the bicycle market. One could say that other bicycle types were eliminated by a kind of natural selection. In other words, the bicycle evolved. Aircraft, communications technology, military technology, and agricultural practices have evolved also. But no one would say that a bicycle or a telephone or a computer modem was alive.

Finally, one could attempt to define life in terms of the second law of thermodynamics, a well-known principle in physics. According to this law, entropy (disorder) of any isolated system must inevitably increase. For example, a glass containing some ice cubes can be said to be in an ordered state; the glass contains both ice and water. As the ice cubes melt, the system will move toward a state which is more random; the highly ordered crystalline ice will eventually disappear and there will be only one substance in the glass (water) rather than two (water and ice). Attempts have been made to define living systems as those which maintain themselves in states of low entropy. I must admit I have never seen the point of this idea. It is true living systems contain a great deal of order. But living systems exchange both matter and energy with their environments, while an isolated system exchanges neither (in the case of the glass with ice cubes, I was assuming that heat absorbed from the environment was small enough to be neglected). In the case of a living organism, the second law of thermodynamics would not apply unless it were completely isolated from its environment. And, naturally, if it were so isolated, it would soon be dead.

It is true that entropy can be defined for a non-isolated system (i.e., one that interacts with its environment), and that the entropy of living organisms is low. But the same can be said of the earth's weather system. Its entropy decreases whenever water evaporates from the ocean and falls on land as rain. The low entropy of the weather system allows it to expend energy in the form of storms, hurricanes, and tornadoes, just as the low entropy of a big cat will allow it to expend energy in bursts of speed. If it is the low entropy of a lion that allows us to call it "alive," then the weather system is alive too.

The problems associated with finding an adequate definition of "life" are naturally not due to the incompetence of biologists. Life is so complex, it is difficult to define it in a few words. For example, the human body contains nearly three hundred different cells types: muscle cells, skin cells, liver cells, and so on, and there are numerous different subtypes as well. The human cortex contains about 100 million neurons (which is, incidentally, about the same as the number of stars in a spiral galaxy), and a single neuron can have connections to ten thousand others.

Even the simplest bacteria are very complicated organisms, and they are capable of interacting with their environments in complex ways. Though they consist of single cells, they contain a multitude of different kinds of biologically active chemicals, including DNA, RNA, and enzymes. Bacteria are able to mutate and evolve (one of the factors involved in developing resistance to antibiotics). A bacterium can respond to changes in its environment, moving toward areas where there is a greater concentration of nutrients, for example. Many bacteria "mate" by exchanging small pieces of genetic material called plasmids. Whatever life is, it is not something that is simple.

Trying to define "life" in a few succinct words is probably a hopeless endeavor. Life can take on too many different forms. Any adequate definition of life would have to be capable of describing a bacterium, a carnivorous plant, a cat, a dormant spore, or any forms of extraterrestrial life that we might some day discover. At present, no one knows of any way to create such a definition.

What Is Life?

Perhaps "What is life?" should be regarded, not as one question, but as many. First, there is the question of the origin of life. We know that life has existed on earth for more than 3.5 billion years. Fossil bacteria have been found which are at least that old, and there is chemical evidence which seems to indicate life existed even earlier. And yet no one is sure exactly how life began. There must have been some kind of chemical evolution which preceded biological evolution. Something as complex as a bacterium or an alga does not spring into existence out of nothing. It seems apparent the chemical systems that led up to the origin of life became progressively more complex, until they reached the point where, suddenly, they were "alive." Or at least this must have happened in one case. Most likely, all life on earth is descended from a single living organism. Unfortunately, we cannot travel 3.5 billion years back in time to observe what was happening. As a result, there are many theories of the origin of life. No one knows which, if any, of them is most likely to be correct.

There are questions relating to evolution. Life evolves by natural selection. All the members of a given species are not genetically identical. Those better adapted to their environment are more likely to survive, reproduce, and pass on their traits to the next generation. Evolution is a never-ending process. Natural selection can take place when there are no mutations. One could observe this simply by making a culture of antibiotic-resistant and nonresistant bacteria of the same species. If some appropriate antibacterial drug were then injected into the culture only one variety would survive. It is true that some of the nonresistant bacteria might develop resistance, but that would be the result of natural selection too.

Natural selection alone would not cause new species to evolve. Something else is needed. That something is mutation, small changes in an organism's genes that will cause its physical form or behavior to change in one way or another. Most mutations are deleterious, or at best neutral. In the former case, natural selection will cause them to be eliminated from a population. But some mutations are beneficial; they cause organisms to be better adapted to their environment. These mutations will tend to be preserved, and will spread throughout a population.

Naturally one mutation, or a few, is not very likely to lead to the evolution of a new species. However, over long periods of time, mutations will accumulate and new species will be born. Plenty of time has been available, of course. There is evidence that a species can evolve significantly in a time as short as 100 thousand years. This is not a long time compared to the billions of years that evolution has been taking place.

Darwin's theory of evolution by natural selection is one of the most significant ideas in all of science. There is so much supporting evidence that it is not possible to doubt its validity. Yet the theory raises as many questions as it answers, questions about the nature of life. One of the reasons we can't adequately answer "What is life?" is that there are still puzzles about the manner in which it evolved. For example, does evolution, as some biologists claim, proceed in spurts? As we shall see in Chapter 5, some scientists think evolution consists of short periods of rapid change interspersed with long periods of stasis. New species may appear suddenly, and then remain more or less the same for periods of millions of years. This claim is disputed by other scientists, who say the change may not be as rapid as it appears to be. For example, sudden changes in the fossil record may appear simply because some species migrated into the area in which the fossils were found. This would look like the sudden appearance of a "new" species. In any case, they add, what is being discussed is change over periods of 50,000 or 100,000 years at the very least. Although such times are "short" in geological terms, they are certainly long compared to the life span of any known organism.

There is the question of whether natural selection explains everything. Few reputable scientists would dispute the idea that it is the most important factor in evolution. But they ask whether other processes might not be involved. For example, are there certain natural forms that evolution tends to produce, or could it tinker together almost anything? Obviously, evolution could not produce an elephant with spindly legs, or a flying hippopotamus. Living organisms are, after all, subject to the laws of physics. Thin legs could not support an elephant's weight; it would break a leg every time it took a step. A flying hippopotamus is a little more plausible, but not very. Such a creature would have to evolve into something very much unlike a hippopotamus before flight became a remote possibility. It is possible to imagine that natural selection might observe certain as yet undiscovered laws of form. It might have a tendency to fit its creatures into certain preexisting "molds."

I should probably point out that this idea is decidedly not part of orthodox Darwinism. Yet there are several points in its favor. One of them is the existence of "parallel evolution" in widely-separated lineages. For example, the Tasmanian wolf, an animal which lived in Australia and New Guinea, and which became extinct early in this century, bears a striking resemblance to the placental wolf, the animal that we all know. The Tasmanian wolf was a marsupial mammal; that is, it carried its young in a pouch. Except for this characteristic, it was hardly distinguishable from its placental counterpart. Not only did it look like a wolf, there were similarities in the shapes of its teeth, claws, ears, and skull. And yet, these two animals were separated by 100 million years of evolution. Both evolved from non-wolflike animals. Was something other than natural selection at work here? Most biologists would say no. There is a minority that believes we can't be so sure about this. Could natural selection alone, they ask, cause the two animals to be so much alike?

And then there is the question of whether living systems may not have self-organizing properties. This is a somewhat more complex question than the previous ones. It has implications that go beyond the field of biology. Hence, I will devote an entire section to it below.

Complexity versus Reductionism, or What Happens when You Pull the Plug in the Bathtub

When you pull the plug from a bathtub full of water, a vortex will form. The water will go swirling down the drain in either a clockwise or a counterclockwise direction. Many people erroneously believe the water will swirl in one direction (don't ask me which) in the northern hemisphere and the opposite direction in the southern. This is simply not true. The coriolis force—one of the forces created by the rotation of the earth—is far too weak to have such an effect. Any small perturbation is sufficient to send the water moving in either direction.

This phenomenon cannot be explained by invoking data about the properties of the water molecule. If all we knew were the size and shape of the molecule, its molecular weight, and so on, and so on, the creation of vortices could not be predicted. I don't mean to imply that physicists do not understand why vortices are formed. They are described perfectly well by the laws of fluid mechanics. I am simply saying this is a case where the technique of reductionism does not work. The behavior of flowing water cannot be explained in terms of the properties of water's basic components.

In some quarters nowadays, it is fashionable to use such words as "reductionist" or "reductionism" as terms of reproach. But the contrary should be the case. It is reductionism that has made modern science successful. A reductionist approach has allowed physicists to delve into the nature of matter. Where, a century ago, some prominent scientists still doubted that atoms really existed, scientists now understand, not only the properties of atoms, but also of their components: the proton, the electron, and the neutron. Protons and neutrons have been further analyzed into their components: the quarks. The theories describing these particles have been confirmed by experiment to a very high degree of precision. It is accurate to say that physicists are approaching a complete understanding of the fundamental nature of matter.

Meanwhile, astronomers, astrophysicists, and cosmologists have used reductionist approaches to gain an understanding of the origin, evolution, and present-day structure of the universe. It is even possible to say what the universe was like shortly after the big bang, when it was only a tiny fraction of a second old. The structure and evolution of stars are well understood, and it is possible to explain the formation of galaxies, the behavior of quasars, which were scattered throughout the universe a few billion years after it was created, and the properties of black holes.

Reductionism has had great successes in biological fields too. It was the development of mathematical genetics during the early years of this century which convinced the last doubters within the scientific community of the validity of Darwin's theory of evolution. More recently, reductionist techniques have uncovered the source of the genetic code, and microbiologists have attained a very good understanding of the workings of plant and animal cells.

However, reductionism does not always work. The formation of a vortex in a bathtub is a good example. Quantum mechanics explains the behavior of the hydrogen and oxygen atoms that compose water molecules. Quantum mechanics also allows scientists to calculate the properties of a water molecule, including its size and shape, and to understand the forces that cause water molecules to be attracted to one another.

Still, quantum mechanics says nothing about the tendency of water to swirl when it is let out of a bathtub. Similarly, the law of gravity says nothing about the reasons why some galaxies have the form of spirals. Astrophysicists understand the motion of spiral arms, and they know that this motion can lead to the formation of pressure waves which induce the formation of stars. But a reductionist approach will not tell them why spiral galaxies form in the first place.

Vortices in bathtubs and the spiral structure of galaxies are examples of emergent properties. They are not created by external forces acting upon the water in a tub, or an agglomeration of stars. They are created by the systems themselves. Another way of saying the same thing is to state that complex systems often exhibit self-organizing properties. "Emergence" and "self-organization" mean more or less the same thing when the words are used in a scientific sense. One could say that organization suddenly emerges when a system is sufficiently complex.

It is not difficult to find nonscientific examples of emergence. It frequently appears in games created by human beings. For example, both chess and checkers make use of small numbers of relatively simple rules. These rules then give rise to games that are complex indeed. A chess master must be aware of many subtle rules of strategy, none of which seem to be a direct consequence of the rules of the game. For that matter, one can observe the same thing in baseball. The rulebook says nothing about the suicide squeeze play or the hit and run.

Emergence is also seen in human societies. However much one knows about human psychology, or the behavior of the individuals in small social groups, that knowledge cannot be used to predict such phenomena as mass media advertising, primary elections, urban blight, or the game of football. We find once again that the whole (the totality of human behavior) cannot be reduced to, or explained in terms of, the sum of its parts.

Naturally it is possible to find numerous examples of emergence in the biological world also. Some species of termite build wedge-shaped nests which are always oriented in a north–south direction. Yet it would be absurd to claim that any individual termite knows what "north" or "south" is. A termite exhibits simple, genetically determined, "hard-wired" behavior, and is only able to communicate with other termites in a limited number of ways. The rules that govern termite behavior are almost as simple as those of checkers. But this doesn't stop termites from building covered runways from their nests to food sources. Some even construct arch-like structures. It appears that termites learned to build arches many millions of years before human beings did.

I could easily give a number of other similar examples, such as the behavior of colonies of ants or bees, or the tendency of certain mammals to form herds. However, I don't want to create the impression that emergent behavior is something only seen in groups of animals. Therefore it might be useful to look at emergent behaviors below the level of the individual organism. The organization of networks of genes provides just such an example.

A human being has about 100,000 genes. Only in rare cases does a single gene, or a gene mutation, express itself by creating a particular bodily trait (such as hemophilia). It is now known that genes operate as components of networks. An individual gene may either inhibit the action of another gene or cause it to be more likely to express itself. Genes are codes for the manufacture of proteins. Proteins interact with one another, and with RNA, in a number of different ways. These interactions sometimes have surprising effects. For example, as Charles Darwin noted long ago, white, blue-eyed cats tend to be deaf. This happens because most genes are members of more than one network. It may perform a function in one that is not related to its function in another. Thus, acting as a part of different networks, a single gene may be associated with fur and eye color, and with deafness.

The interaction of genes is not yet very well understood. However, one thing is clear. A knowledge of the structure of DNA (genes are essentially pieces of DNA) will never be sufficient to understand gene expression within the human body. There is much to be learned about gene expression and interaction. It is clear this is not an area in which a reductionist approach is likely to work.

What Is Complexity?

What, exactly, is complexity? This is not an easy question to answer. Complexity is as hard to define as life. Perhaps this is not so in the world of mathematics. Mathematicians have spent a great deal of effort working out ways to speak about the complexity of a number or the complexity of a computer program. But their work has little applicability to the real world. For example, the computability of a number seems to have little to do with the complexity of the genetic makeup of a dog or the complexity of an ecosystem.

We can generally distinguish between complex systems and simple ones. For example, everyone would agree that a chicken is complex, while a plastic marble is not (at least I know of no self-organizing properties exhibited by the latter). A tub full of water probably lies somewhere near the borderline. After all, it is made up of molecules which are all exactly alike (we can ignore the possible presence of soap or hair, which has no effect on vortex formation). Some of the water molecules may possess more energy than others. This does not significantly affect the behavior of the tubful as a whole.

Strangely, the difficulties associated with defining "complexity" do not have any important consequences. Just as biologists are able to study life without coming to any final agreement as to exactly what life is, it is possible to forget about definitions and to study the self-organizing properties of complex systems without any qualms.

It is possible to devise measures of complexity that apply to certain specific cases. One could, for example, equate the complexity of an ecosystem to the number of plant and animal species that inhabit it. Better yet, one could make an estimate of the number of different interactions between the species. Similarly, it is obvious that a human body, which contains about 100,000 genes, is more complex than a bacterium, which has only a thousand. The number of genes may or may not be the best way to define the complexity of an organism, because gene networks are left out of the picture. However, counting them can at least give us an idea of how complex an organism they are likely to produce.

Perhaps there is no obvious way to compare the complexity of an organism to that of an ecosystem, a traffic jam, or even a market in pork bellies. However, this is of little importance. One generally doesn't compare *E. coli* bacteria to pork bellies anyway. There are cases where quantities must be precisely defined. It would be impossible to talk about magnetism without defining the strength of a magnetic field. But there are other areas in which the lack of an all-embracing definition has proved to be no hindrance.

Compute, Compute, Compute

I am writing this book on a desktop computer, which is far more powerful than the mainframe computer I used when I was a graduate student. It has much more memory than the mainframe had, and it is many times faster. The equations that arose out of the research I did for my doctoral dissertation could now be solved much more easily and quickly than was possible then (and I don't have to contend with the antics of engineering students, whose programs for drawing naked women exhausted the storage capacity of the mainframe's hard disk). Finding space for programs and data is not the problem it once was. Personal computers are nowadays so powerful they are often used for scientific research that, at one time, would have required the largest and fastest computers in existence. Most of the work in the sciences of complexity can be performed on computers which are slower and less powerful than those required in order to play the newest computer games. This observation may or may not be a comment on contemporary life.

The advent of modern, high-speed computers has affected virtually every field of scientific endeavor. Research in high-energy particle physics would be impossible without computers. Computers are used in astronomy, biochemistry, aerodynamics, and meteorology, to name just a few fields. They have even affected the practice of pure mathematics—mathematics which is carried on for its own sake, without giving any thought to possible applications. Mathematicians who once worked with pencil and paper are now using computers to help them prove complex theorems. Paleontology, which concerns itself with the classification of fossil species, is one of the few fields not affected much by computers. Even this is certain to change.

Finally, computers have made the development of new scientific fields possible. It is impossible to imagine doing research in artificial intelligence without them. And it is the computer which has given rise to the sciences of complexity, which constitute the subject matter of this book.

It is the computer which has provided an answer to the question, "How does one go about studying a complex system?" Recall that complex systems cannot be approached in the time-honored reductionist fashion. Emergent properties typically cannot be analyzed in terms of the system's components. It is often difficult or impossible to write down equations which describe self-organizing behavior. When they can be written, they generally turn out to be too complicated to solve.

A modern computer, which can handle large quantities of data and perform calculations at high speed, can be used to analyze self-organizing systems in a different way. If it is impossible to construct a simple theoretical model of a system, one can create a simulation of such a system inside a computer. The techniques involved are not unlike those employed by engineers, who will create a computer simulation of an aircraft before a prototype is built, or of automotive engineers, who create simulations of new models. The difference is that scientists who do research in the sciences of complexity are trying to answer fundamental questions such as: "How did life begin?" "How did it evolve?" and "What brought about the evolution of complexity in living organisms?" The last question is, incidentally, an especially interesting one. For the first 3 billion years in the history of life on earth, only single-celled organisms existed. Then, suddenly, multicellular life evolved. What brought about this sudden increase in complexity? Could it have happened much earlier? Or not at all? As we shall see, there are a number of differing opinions on these points. There are even some scientists who deny there was any great increase in complexity at all. At the moment, it is not possible to answer the questions raised by the evolution of complex life forms; this is a question I will discuss over the course of several chapters. But it is safe to say that, if any answers are eventually discovered, it will be with the help of the computer.

There is a lot of controversy surrounding the sciences of complexity. Most of it centers around the practice of creating computer models. The critics suggest that scientists who concern themselves with complexity are doing nothing more than playing elaborate mathematical games. There is no guarantee, they say, that computer models have anything to do with reality. For example, if evolution is modeled on a computer, the results may be interesting, but there is no particular reason to think that this tells us anything about evolution in the real world. If a colony of simulated ants is created, they may or may not interact with one another in the same ways real ants do. Even if the behavior of the simulated colony resembles the behavior of a real one, this is no guarantee that both operate in the same way. It is entirely possible that using different sets of rules for the behavior of individuals may have more or less the same effect.

This can be a telling point. Certainly, an aircraft model can be modeled on a computer. The results can be quite useful to the aircraft engineers working on the design. But then human beings have a great deal of experience with building airplanes. They have been at it for almost a century. On the other hand, a computer model of gene interactions may or may not tell us anything significant. There is generally no way to check the results against interactions in the real world. These are not well enough understood.

It is these very things which make research in the sciences of complexity seem have so much promise. We don't have a deep understanding of the interaction of genes, so the use of computer models could very well give us new insights. No one knows precisely how life began on earth. Again, those working in the field of complexity may give us some exciting new ideas. The engineers who model aircraft generally don't gain any significant new insights into the phenomena of aerodynamics. This is a field that has been well explored. Research into the nature of life could very easily expand, because there is so much that we do not know.

The Origin of Life

As I have already pointed out, no one really understands how life began. However, as we shall see in Chapter 3, theoretical biologist Stuart Kauffman has an idea about the origin of life, one he finds so compelling he cannot imagine life starting in any other way. Much of Kauffman's work is highly mathematical in nature. Even professional biologists sometimes find it hard to understand. At the root of his work, there is a simple idea: chemical systems can evolve toward a greater degree of complexity; once a certain level of complexity is reached, life will spring into existence. According to Kauffman, life is an emergent property of a certain kind of complex system.

Traditional scientific speculation about the origin of life has centered upon interactions between RNA and proteins. No one seriously believes life began with the creation of DNA, which carries the genetic code in all living creatures today, because DNA cannot replicate itself without the aid of enzymes. Without DNA, these enzymes cannot be manufactured in the first place. Recently, some Yale University scientists have discovered that some DNA can exhibit enzyme action. It is not clear what implications this has for the problem of the origin of life. They did not find that DNA could replicate itself, only that certain kinds could snip themselves in two. At present, there are basically two contending theories (and a lot of variations and subvariations) about how life originated. One theory says proteins came first, and RNA evolved later. The other theory states just the opposite, that RNA was created first. Both theories depend upon the idea that chance combinations of organic chemicals could lead to the formation of the molecules that form the basis of life.

In Kauffman's theory, it doesn't matter which molecule came first. I'll be going into the details of his ideas later. It seems that if some way is found to confirm them experimentally, they will be established as an important conceptual advance. We will then be several steps closer to being able to answer the question, "What is life?"

Life in a Computer

The field of artificial life is one in which a great deal of innovative research is currently being carried out, some of it theoretical in nature. Kauffman's theories have been influential, and he does make use of computers in his work. But many scientists want to pursue more practical ends. They look at the results produced in "real" computer simulations. Pondering questions about the nature of life has led some of them to ask, "If one wants to study life, why not create it?" Naturally they are talking about creating artificial life forms that live in computers.

Ecologist Tom Ray, for example, has created an artificial world named "Tierra" that exists inside a computer. The electronic computer-virus-like creatures that inhabit this world reproduce, mutate, evolve, and die. Though their genetic code is very simple—Ray seeded his world with electronic organisms possessing only three genes, which served no function but to allow his electronic organisms to reproduce—the simulation has given rise to astonishing evolutionary results. For example, the first time Ray ran his experiment he was able to observe the evolution of parasites. He next saw the hosts develop a defense against parasites, driving the latter to extinction. But then a new kind of parasite evolved, and Ray's electronic evolutionary arms race went on.

As I write this, Ray is completing a project which will allow electronic life to migrate back and forth between various computers connected to the Internet. He hopes the more complex environment will allow him to observe phenomena not seen in the simpler version of the experiment. In particular, he wants to see if he will observe something analogous to the great increase in biological diversity that followed the evolution of multicellular life on earth.

Do Ray's electronic organisms deserve to be called alive? According to some definitions of life they do. It would be fruitless to argue the question at length. The important thing is that Tierra and its successor, called "network Tierra," allow scientists to observe evolution in a computer. At present there is no way to compare Ray's results with evolution in the real world. The experiments are certainly likely to provide some interesting insights, and to give rise to electronic evolutionary diversity as well.

Complexity in the Laboratory

"Experiments" in the sciences of complexity are usually performed on computers, dealing with questions not answered by laboratory research. Therefore, accounts of such research generally place a great deal of emphasis on the results of computer "experiments." However, it is also possible to study complex systems in the laboratory. At the Scripps Research Institute in La Jolla, two scientists are pursuing investigations that are related to the origin of life. Julius Rebek Jr. has created synthetic chemicals which have some of the characteristics of living systems. His colleague, Reza Ghadiri, working in a different laboratory, has been experimenting with biological chemicals. Among other things, Ghadiri has created a small protein which is capable of replicating itself. Replication is one of the characteristics of living organisms, thus Ghadiri, too, has taken steps toward the creation of life.

Neither Rebek nor Ghadiri is trying to discover how life began on earth. Ghadiri stresses that he does not know how life began and professes not to care. Neither Rebek's nor Ghadiri's work bears much resemblance to traditional research into the origins of life. The latter research has never lead to any definite conclusions. If it had, there would not be so many theories about life's beginning.

In the past, scientists who have wanted to delve into the origin of life have tried to recreate the conditions that are thought to have existed on the primordial earth. They created chemical "soups" containing such biological chemicals as RNA and proteins and attempted to see if this would reproduce some of the steps that eventually led to the evolution of the first living organisms. Rebek and Ghadiri, on the other hand, are experimenting with simple chemical systems in order to understand the ways in which complexity may arise.

Rebek doesn't speak much of complexity. But it is something that is very much on Ghadiri's mind. He points out that complexity may arise in a lot of different ways, and he wants to know what some of them are. Though the creation of proteins that can reproduce themselves (Rebek's synthetic chemicals, incidentally, can reproduce too) doesn't allow us to understand how life began, it does give us new insights into the ways in which complex biological chemicals can interact.

New Horizons or Empty Promises?

In one sense, the last few sections have been a preview of some of the research that will be discussed in this book. The sciences of complexity, as young as they are, have produced a great number of new, original ideas, and I wanted to cite a few examples before discussing the topic in more detail. At the moment, it is difficult to tell which, if any, of these ideas is likely to lead to the discovery of important new knowledge. After all, a concept can be new, original, and exciting, and yet turn out to be wrong.

One can't help but think of the field of artificial intelligence, which was also an area of research that seemed full of promise. Though research has been going on for more than two decades, much of its promise remains unfulfilled. True, some important results have been obtained, and some striking advances have been seen. Don't be too impressed by such highly publicized results as the defeat of world chess champion Garry Kasparov by an IBM computer. The computer won because it excelled at "brute force" calculation. Its "understanding" of strategic themes on chess was comparatively poor. It certainly didn't "think" about chess the way a human being would, and the expertise it exhibited mimicked only one facet of human intelligence.

Workers in the field have not achieved nearly as much progress as they thought they would when it was just beginning. I think that researchers in the science of complexity will achieve much more, in a shorter period of time. Work on complexity is motivated by one basic theme: complex systems have self-organizing properties, and these properties can be fruitfully studied. Artificial intelligence, on the other hand, was based on many ideas. Research in the area never had the direction that the sciences of complexity have already.

Furthermore, most of the things we observe in nature are complex systems, while only some of them exhibit intelligence. Real simplicity is something that is relatively rare. In order to see this is so, look at physics, the reductionist science par excellence. Physics has been so successful because physicists have succeeded in breaking systems down into their components and in understanding the behavior of the basic constituents of matter.

Physics has sometimes been called "the science of approximations." What this means is that most systems are not simple enough to be described exactly. For example, one can use Newton's law of gravity and his laws of motion to describe the manner in which two gravitating bodies revolve around one another. As soon as you add a third body, the equations become too complex to be solved. To be sure, accurate approximate solutions can be obtained. One can calculate the date and time that an asteroid will pass by the earth fifteen or twenty years in the future, even though the asteroid is subject to many gravitational influences, including those of the sun, moon, and various planets. However, even gravitational problems become intractable when they are too complex. Astronomers don't know whether or not the solar system is stable. The nineteenth century French mathematical physicist Pierre Simon de Laplace thought that he had proved that it was, and for more than a century scientists accepted his results. However, recent work in chaos theory suggests that, over long periods of time, this may not be true. For all we know, at some point in time Mars may go bounding out of the solar system.

Similarly, scientists have no trouble describing the behavior of the simplest atom, hydrogen. The equations obtained can be solved exactly. A hydrogen atom is made up of only two particles: a proton, and an electron that orbits around it. As soon as a third particle is added, exact solutions no longer exist. A helium atom consists of two electrons and a nucleus. Its behavior can be predicted to a very high degree of accuracy. Again, the solutions are approximate.

If a helium atom can be described only in an approximate manner, imagine the problems associated with trying to use reductionist techniques to determine how a very complex molecule such as a protein will behave. Clearly, new approaches are needed. Those new approaches are currently being sought, and sometimes found.

Laws of Complexity

One interesting question is whether there are any "laws" of complexity which might be analogous to, say, Newton's laws of motion. That is, are we likely to discover any principles that will apply to complex systems in general? Or will it be necessary to study each type of system individually? I would incline toward the latter view. I have always been suspicious of theories that try to be too embracing. I could turn out to be wrong. That thought creates a sense of anticipation in me. I am anxious to see what is going to be discovered.

If I turn out to be wrong about general laws of complexity, this will not bother me. I'm interested in broader issues. I first became interested in physics, not because I desired to know what Newton's laws of motion said, or what Schrödinger's equation meant, but because I wanted to find out what science could tell us about the fundamental nature of physical reality. Similarly, I became interested in the sciences of complexity because I wanted to know what light they could shed on that multifaceted question, "What is life?"

2

CREATING LIFE
IN THE LABORATORY

Theoretical biologist Stuart Kauffman thinks that he knows how life began. Working with computer simulations at the Santa Fe Institute in New Mexico, he has developed a plausible scenario for the origin of life. According to Kauffman, it doesn't matter whether life began with RNA, proteins, or with some other organic chemicals. All that is needed is a certain degree of complexity. Once this is attained, life will spontaneously emerge.

Where Kauffman views the problem of creating living organisms in the laboratory in purely theoretical terms, Julius Rebek, however, really is attempting to create life, or at the least understand the steps that might have led to its origin. He has crafted synthetic chemicals which mimic several of the characteristics of living organisms. They can reproduce, mutate, and evolve, and they enter into symbiotic relationships with other reproducing chemicals.

M. Reza Ghadiri has a goal that is hardly a modest one. He would like to understand the nature of "life itself," gain an understanding of the origin of life, and learn how to create it. Unlike Rebek, he does not work with synthetic chemicals. Instead, he uses self-replicating peptides, that is, small protein fragments. His peptides can catalyze the formation of other peptides, form symbiotic relationships, and spontaneously develop error-correcting mechanisms that prevent mutant forms from proliferating.

Thus we have three different new and original approaches to the origin of life: a theoretical one and two different experimental ones. All three are based on the idea that life is an emergent property that arises in certain kinds of complex systems. I will be discussing the work of Kauffman, Rebek, and Ghadiri in detail in this chapter. Before I do, it is necessary to cover some of the characteristics of living organisms and look at some of the theories that have been advanced to explain their origin.

The Origin of Life

Various kinds of scientific evidence indicate that life sprang into existence with extraordinary rapidity once the terrestrial environment became hospitable. Early in its history, the earth was bombarded by giant asteroids and meteorites. It may even have been struck by a body as large as Mars. The impacts created so much heat that the surface of the earth remained molten for hundreds of millions of years after it was created. Many of the objects that struck the earth at this time released many times more energy than the impact of the asteroid that led to the demise of the dinosaurs 65 million years ago. The bombardment may have continued for as long as 800 million years, from the time that the earth was formed 4.6 billion years ago, to a time about 3.8 billion years before the present. Any living forms created could not possibly have survived in such an environment. Yet a billion years after earth was formed, it teemed with life. Fossils of organisms resembling blue-green algae have been found in 3.5 billion-year-old rocks in Australia and Africa. Chemical traces of life dating back to 3.85 billion years have been discovered in rocks found in Greenland. Of course if living organisms did exist 3.85 billion years ago, they are not necessarily our ancestors. They could have been wiped out by an impact with some large body, and life could have begun anew later.

There is every reason to think that the ingredients needed for the creation of life were present almost as soon as the earth's crust cooled enough to allow oceans to form. Amino acids, the building blocks from which proteins are made, have been detected in space as have other organic chemicals. They could have been carried to the earth's surface on falling interplanetary dust particles, or they could have fallen from the tails of comets that brushed the earth's atmosphere. Alternatively, they could have been created on the earth itself. Numerous experiments have shown that this is possible.

The first of these experiments was performed in 1953 by Stanley Miller, who was then a graduate student at the University of Chicago. At the time, it was thought that the primeval earth's atmosphere consisted of hydrogen, methane, and ammonia. Miller put these gasses in a sealed flask. This flask was connected to another flask containing water. Miller heated the latter flask so that water vapor would be present. He then passed an electrical discharge through his mixture. The idea was to imitate conditions on the early earth. The flask of water was a simulated "ocean," and the electrical discharge was analogous to lightning. Miller let his experiment run for a week. At the end of that time, he found that he had created significant quantities of organic compounds, including two of the amino acids that are the components of the proteins found in living organisms.

Today, scientists no longer believe that hydrogen, methane, and ammonia were major components of the earth's atmosphere at the time that life originally evolved. They think that the atmosphere was made up primarily of carbon dioxide and nitrogen, and that methane and ammonia were present only in small amounts. However, experiments like Miller's have been performed in a number of different simulated atmospheres. In many of them, amino acids, as well as nucleotides—the components of DNA—have been formed. It is somewhat more difficult to create nucleotides than amino acids, but it can be successfully done.

Amino acids and other chemicals necessary to life will not form if oxygen is present. Scientists believe that this gas was not present on the earth 3.5 or 4 billion years ago. It is thought that all, or nearly all, of the oxygen in our atmosphere was created by organisms capable of carrying out the process of photosynthesis, and that they began to release oxygen in large quantities only about 2 billion years ago. Thus there is every reason to think that the experiments simulate the atmosphere of the early earth in a reasonable way.

The building blocks of life were present in the primeval oceans. Yet life could not have originated there. Amino acids and other chemicals were only present in relatively small quantities. As a result, the chemical reactions which would have created more complex molecules could not have taken place often. In such a dilute mixture, the basic components of life would rarely encounter each other. Furthermore, water tends to break linked amino acids apart. If any of the long chains of amino acids that we call proteins had been created, they would have disintegrated almost immediately.

However, there did exist environments where the chemical reactions that led to the creation of the first living organism would have taken place more often. If the sun's heat caused the water in a tide pool to evaporate, the solution of organic chemicals would have become more concentrated, and reactions would have taken place more frequently. Another currently popular theory pictures the chemical evolution that led to life as taking place on clay substrates. The idea is that if chemicals adhere to a two-dimensional clay surface, they will "meet up" with one another more often than they will in a three-dimensional pool. It is not hard to see why this should be the case. It is a lot easier to bump into someone on a two-dimensional sidewalk than it is to collide with a bird while skydiving.

Is Creating Life Really That Easy?

The current scientific conception of the origin of life can be summarized as follows. As the building blocks of life were washed up from the oceans, they were concentrated in evaporating tide pools or on clay surfaces. More complex chemicals began to form. Some of these randomly produced chemicals developed the ability to replicate themselves, using simpler chemicals present in their surroundings as building materials. Chemicals best at reproducing themselves quickly became the most numerous. When one of these self-replicating complexes learned to enclose itself in a membrane, the first living cell came into existence. The last step, incidentally, does not seem to be an extremely difficult one. There are organic substances that spontaneously form tiny spherical bubbles, and some of the replicating chemicals could have become trapped inside them on one or more occasions. The membranes of our cells, for example, are made of chemicals called lipids that are formed from fatty acids. Lipids can be made to curl up into enclosed vesicles when they are placed in water. Finally, such a complex of chemicals could reproduce itself if the bubble in which it was enclosed split in two when it reached a certain size.

This simplified description of the origin of life makes it seem a simple and inevitable process. In reality, it was not so easy. There is much about the evolution of the first living organisms that scientists do not understand. There are several competing theories, and all of them are speculative to some extent.

To understand the difficulties that scientists face, it is necessary to know something about the manner in which cells make a living. All cells make use of a genetic code that is stored in molecules of DNA (deoxyribonucleic acid). Two strands of DNA twist around one another like two intertwined springs, forming a structure known as the double helix. Each strand of DNA is constructed from four nucleotides called adenine, guanine, cytosine, and thymine (abbreviated with the capital letters A, G, C, and T) and chemicals that bind the nucleotides together in long chains.

If a cell is to survive, it must manufacture proteins. This is done in the following manner: First, an enzyme (an enzyme is just a special-purpose protein) separates the two strands of a section of the DNA double helix. Then, a strand of RNA (ribonucleic acid) is formed on the DNA template. RNA is also made up of nucleotides. These are A, G, C, and U (uracil); in RNA, U is substituted for the T in DNA. This process transfers the genetic information from the DNA to the RNA. The cell then uses the RNA to manufacture proteins.

The genetic information in both DNA and RNA is encoded in sets of three nucleotides, called triplets. AGC or CGU are two examples. There are four nucleotides, thus there are sixty-four possible combinations. There are four different "choices" for the first nucleotide (for example, in DNA, the choices are A, G, C, and T), four different choices for the second, and four for the third. A little simple arithmetic gives the result: $4 \times 4 \times 4 = 64$. Most of the sixty-four triplets code for the formation of a specific amino acid. There are also three that function as "stop signs." They mark the point at which the reading of the DNA is to stop. There is a certain amount of redundancy in this arrangement. Although there are sixty-four triplets, living organisms make use only of twenty amino acids. Typically, several different triplets will code for the same amino acid. For example, the triplet GUC in a RNA strand tells the cell to make the amino acid called valine. So do GUU, GUA, and GUG.

Suppose that the cell's machinery finds GUC at the beginning of an RNA strand. It will then synthesize valine. Going on to the next triplet, it will make another amino acid that is linked to the first. Eventually, it will come to the "stop sign," which tells the cell that the synthesis of the protein is complete. Thus the reading of a series of RNA triplets leads directly to the formation of a protein, which is a long chain that is typically made up of several hundred amino acids. This process can be summarized as follows:

$$DNA \rightarrow RNA \rightarrow protein$$

I have omitted some of the details of this process to emphasize the essential way the information flow operates. Because I am stressing the basic ideas involved, I do not think it necessary to describe how proteins are made. I should mention, however, that what I am calling "RNA" is generally referred to as messenger RNA (or mRNA). Other kinds of RNA are also found within a cell. They perform different functions.

There is one important point to note. A protein enzyme is needed to split the two strands of DNA apart. Enzymes are thus needed if DNA is to replicate itself, because replication cannot take place without the splitting-apart of the helical strands. This implies that the first living cells could not have used DNA to carry their genetic code. If DNA somehow evolved, it would not have been able to reproduce without protein helpers. And it could not have made protein helpers by itself.

Scientists therefore believe that the first living organisms used RNA, or some precursor of RNA, to carry their genetic code. Presumably, when the more stable DNA evolved, cells with protein enzymes already existed. DNA replaced RNA as the carrier of the genetic code, and RNA was relegated to other tasks.

Which Came First: RNA or Proteins?

Early living cells contained RNA and protein. But which came first? If scientists could answer that question, they would have gone a long way toward understanding the origin of life. At present, a number of theories attempt to deal with this matter. According to one, life began with the evolution of "naked genes" consisting of strands of RNA. Another theory says proteins evolved first. And yet another theory maintains life evolved not once, but twice. That is, protein organisms and RNA organisms evolved separately, and then entered into a symbiotic relationship with one another.

The RNA theory is the most popular. In 1983, Thomas R. Cech of the University of Colorado and Sidney Altman of Yale University independently discovered that certain kinds of RNA could act as enzymes. It had previously been thought that all enzymes were proteins. This discovery gave rise to the idea that ancient RNA molecules could have brought about the replication of other RNA molecules. If this was the case, the RNA could have reproduced, mutated, evolved, and eventually have invented protein synthesis.

Although in some respects this is an attractive theory, it meets with a number of difficulties. First, when scientists set up experiments in which they try to duplicate the conditions that might have existed on primeval earth, they find it difficult to synthesize the nucleotide components of RNA. It is much easier to create amino acids. Furthermore, it appears that if self-replicating RNA had evolved, it would have mutated much too rapidly. When modern DNA is replicated, errors frequently appear. For example, a G may appear where a C should be. These are corrected by "proofreading" enzymes which snip out the incorrect pieces of DNA, and allow them to be replaced by the proper nucleotide sequences. But if life began with RNA, there would have been no proofreading enzymes.

Errors would have accumulated, and, most likely, the RNA code would have quickly turned into nonsense.

Thus some scientists think that proteins came first. This theory meets with difficulties, too. In the absence of any genetic code, primitive "cells" would not have been able to pass on their characteristics to their descendants. If only proteins existed, there could have been metabolism, but there probably would not have been any heredity.

Naturally it is conceivable that life arose twice, and that self-replicating RNA and proteins formed a symbiotic relationship with one another. This idea meets up with the same difficulties encountered by the RNA theory. The double-origin theory doesn't make the evolution of the first living cells seem any easier.

The most radical theory of the origin of life is one proposed by the Scottish chemist A. G. Cairns–Smith. According to Cairns–Smith, the first replicators might have been clays. He notes that clays and muds are made up of crystals, and suggests that the organic molecules deposited on the clays might have been lined up in ways that imitated the crystalline structure. Cairns–Smith's theory is not the same as the hypothesis that organic chemicals became more concentrated when they were deposited on clay surfaces. In Cairns–Smith's theory, the clays play a truly active role.

Although crystals have regular structures, they can form patterns that are very complicated. This is caused by the presence of flaws. Every now and then some of the crystal's atoms will be misplaced, or one type of atom will be replaced by another. This gives rise to intricate patterns that can reproduce themselves. For example, when a second layer of clay is deposited on the first, it will imitate the same pattern. According to Cairns–Smith, life began with self-replicating clays. There would be a kind of evolution because some structures would replicate themselves more efficiently than others. Successful clay "life-forms" could spread. For example, if a clay dried out, the wind might blow particles into another moist environment where it could replicate again. This theory suggests that organic life began when some of the organic chemicals that became attached to the crystals began to reproduce at a rate that far exceeded that of their hosts. If Cairns–Smith is correct, RNA could have existed for a long time before it became self-replicating.

This theory is somewhat implausible. Structures that were complicated enough to produce RNA or some RNA predecessor would almost certainly not have been able to reproduce themselves accurately. Whenever a second layer of clay was deposited on an existing layer, new flaws would be likely to occur. The replication would not be exact, and over time errors would proliferate. There is also the question of whether the patterns in the clay could have been complex enough to give rise to living molecules. However, the theory cannot be completely ruled out, because it has not really been tested in the laboratory.

Creating Life

Stuart Kauffman thinks that it should be possible to create life in the laboratory. "We should be able to create life anew in the fabled test tube," he says. Kauffman has created a theory of the origin of life that is not based on the assumption that either RNA or proteins came first. He thinks that when any collection of biochemical molecules becomes complex enough, it will automatically spring into life. According to Kauffman, these molecules could be peptides (short chains of amino acids), nucleotides, or both. He believes that life is an emergent property of a complex system of molecules.

Kauffman's theory is based on the idea of autocatalytic sets. He notes that most biochemical reactions take place very slowly, but they can be speeded up considerably if a catalyst is present. For example, molecules A and B may link together at a very leisurely pace to form molecule C. But if molecule D is present, they may combine very rapidly. D here is called a catalyst. It remains unchanged as A and B combine. Thus it is able to catalyze other A–B reactions after it has played its role in the first.

Numerous catalytic reactions take place within a living cell. The protein enzymes that act on DNA and RNA can be said to be catalysts. Other enzyme catalysts allow a cell to make use of food energy and to regulate its metabolism in numerous different ways. Thousands of different chemical reactions occur within cells, and they all depend upon enzymes.

An enzyme can speed up a reaction by a factor of 100 million to 100 billion, even if it is only present in small quantities. Enzyme catalysis takes place because proteins and other biological molecules have complex three-dimensional shapes. If two molecules of complimentary shape encounter one another, they will fit themselves together. A simple example would be that of a concave molecule fitting inside one that was convex. Sometimes the complimentarity of the shapes is imperfect. The molecules fit themselves together only with difficulty. Catalysts remedy this situation. An enzyme catalyst may attach itself to two different, smaller, molecules, lining them up so that they fit together, thus encouraging them to combine.

Again, this is a somewhat simplified description. Enzymes often do more than align molecules. They sometimes induce distortions in the shapes of the reacting molecules (A and B in the above example), making them more likely to combine. Some enzymes break molecules apart rather than link them together. We have already encountered an example of this in the discussion of the enzyme that splits apart two intertwined strands of DNA. However, the picture of an enzyme as a molecule that other molecules fit into is a reasonably accurate one.

Cells are very complex entities. Human DNA contains about 100,000 genes. Since genes code for proteins, a human cell contains about that number of different biological chemicals. The simplest known bacterium, called pleuromona, has been estimated to have somewhere between a few hundred and about a thousand genes. It appears that, below a certain level of complexity, life could not exist, and pleuromona may be near that level. It is reasonable to assume that a cell must have a certain degree of complexity. After all, if the strands of DNA were too short, it could not contain enough genes to program a cell's structure, reproduction, and metabolism.

According to Kauffman, cells are autocatalytic. To see what he means, let's look at the analogy he uses. Suppose, Kaufmann says, that you spread a large number of buttons on the floor, and then take out some thread and begin tying them together in pairs in a totally random manner. Any button may be connected to any other. You might, for example, decide what connections to make by flipping coins. If you then pick up a button, you will either find that it is not connected to any others because it was never threaded, or you will find yourself holding a string of interconnected buttons. When you have not threaded a great number of buttons together, you will usually not find yourself picking up very many at a time. If, on the other hand, you have put in a lot of threads, the opposite will be the case; pick up one, and you will find that a lot of others are tied to it. If there are enough threads, it is likely that you will find yourself lifting the entire ensemble of buttons from the floor.

All this seems very obvious. The more threading you do, the more the buttons are interconnected. In a way, it's like municipal transit. The more bus, subway, or train lines there are in a city, the greater the number of points you can get to without walking more than, say, two blocks.

Now, if this were all there was to it, then Kauffman's analogy would be of little relevance. However, there's more. It can be proved mathematically that the interconnectedness of the set of buttons (or the set of destinations that can be reached by bus without too much inconvenience) begins to change dramatically when the number of threads is about half the number of buttons. Once this point is passed, the buttons tend to be linked in large groups where before there were mostly a lot of small networks of them. If there were, say, a thousand buttons on the floor, you would likely find yourself holding a net of, say, eight hundred when you tried to pick one up.

Kauffman likens the set of threads and buttons to the catalytic reactions that take place within a cell. The set of catalyzed reactions is also joined together in large networks. The more catalysts there are, the more likely the chemical reactions that take place within a cell will be interconnected.

Suppose we have a set in which there are ten different kinds of molecules (peptides or nucleotides or both), and that each molecule has a random chance of one in a million of catalyzing one of the reactions that can take place. It's obvious that nothing much will happen. The chances are, ten molecules will not be able to catalyze any of the possible reactions. We will have nothing more than a collection of ten "dead" molecules. This situation is analogous to a set of buttons in which the interconnectedness is low.

As you attempt to follow the ensuing argument, it might be best to remember that Kauffman's analogy, like all analogies, is an imperfect one. In the analogy, there are two kinds of objects: buttons and strings. They are not interrelated. If we have a given number of buttons, we are free to add as many strings as we want. In the case of a system of organic molecules, there is an interrelation. As the number of different kinds of molecules (buttons) increases, so does the number of catalysts (pieces of string). This happens because every molecule is likely to possess the ability to catalyze some reaction or another.

Kauffman notes that, as the number of different molecules increases, the number of reactions increases much more rapidly than the number of molecules. This can be illustrated by the following simple example. Suppose we have two molecules, A and B, and that they can link up to form the pairs AA, BB, AB, and BA (if the molecules are not symmetrical AB and BA will not be the same). Four different reactions can create pairs. These pairs can also break apart, making the total number of possible reactions equal to eight. But if you add one more molecule—call it C—the number of possible reactions increases to eighteen. In this case there are nine possible pairs that can be formed: AA, BB, CC, AB, BA, AC, CA, BC, and CB. There are nine reactions that create them, and nine reactions that can split them apart. Even in this simple case, the effects of increasing the number of elements in the set is quite dramatic. Increasing the number of molecules by 50 percent has increased the number of possible reactions by 125 percent.

Kauffman then points out that, as the number of molecules and the number of reactions increases, the chance that any given molecule will be able to function as a catalyst will increase dramatically. Suppose there are a million reactions. If the molecule has a one-in-a million chance of catalyzing each one, the chances are pretty good it will become a catalyst. A huge network will form in which each chemical has a good chance of catalyzing one reaction or another. In Kauffman's terminology, such a set is said to be autocatalytic.

Admittedly, the figure of one in a million is arbitrary. However, Kauffman is not looking at the behavior of a group of chemicals in quantitative terms. He is trying to understand the logical structure of a biochemical system. The particular number is not important. If the chance were one in two million, for example, the same threshold would be reached when there were two million reactions.

Kauffman contends that autocatalysis is the basis of life. Life appears when a set of interacting chemicals reaches a certain level of interconnectedness. The only other thing needed to bring a living cell into being is some kind of vesicle in which the chemicals can be enclosed. Once this happens, the system should be able to reproduce. By making use of "food" chemicals in its environment, it would be able to grow, and eventually split in two.

Finally, Kauffman says, autocatalytic sets would have the ability to evolve. Spontaneous reactions could create new molecules that were not part of the autocatalytic set. If the set then incorporated some of these molecules into itself, it would mutate into a new, slightly different set. According to Kauffman, it is possible that living entities reproduced and evolved before the genetic code embodied in DNA and RNA came into existence.

Obviously, Kauffman's theory is quite different from those previously discussed. He sees life as a property that emerges out of mathematical necessity when some threshold of molecular diversity has been reached. The theory does not depend on what kinds of molecules were created on the primeval earth. It is only necessary that some of them be capable of catalyzing some reactions. Since there will be some reactions that take place spontaneously, the number of chemicals will gradually increase. When there are enough of them, the requisite degree of diversity will have been created.

The drawback to this theory is that it is hard to see how it could be tested experimentally. As Kauffman himself points out, "No one understands how the complex cellular networks of chemical reactions and their catalysts behave, or what laws might govern their behavior." It is likely to be a difficult or impossible task to create a system in the laboratory that would become autocatalytic at some point. Indeed, the logic of Kauffman's ideas is appealing. However, there have been many beautiful, logical theories that turned out not to be true.

Synthetic Life

Julius Rebek would like to gain some insights into the nature of life. He is attempting to do this by creating synthetic cells which mimic life in a number of important respects. His chemicals can make use of "food" molecules in their environment, replicate themselves, mutate, and evolve. They can also enter into symbiotic relationships with other synthetic chemicals.

To gain an understanding of how life might have begun, Rebek began, in 1990, trying to create molecules that could replicate themselves. He used synthetic, and not biological molecules, because he wanted to gain an understanding of the basic principles of replication. This didn't require real biological reactions. He saw the creation of these chemicals as a first step toward gaining an understanding of life and its origin.

As we have seen, organic molecules catalyze the replication of other molecules by virtue of their shape. Chemical attractions may also be involved. For example, a hydrogen atom in one molecule may link up with an oxygen atom in another. This makes the "fit" even better. After thinking about this, Rebek realized that molecules designed to have certain shapes should be able to replicate themselves. So he set out to see if he could synthesize one or more such molecules.

The first self-replicator Rebek created was a J-shaped molecule called ARNI (adenine ribose naphthalene imide). ARNI assembled copies of itself from chemicals that it found in the liquid solution in which it had been placed. The new ARNI molecule was easily created because it could fit inside the original one if the two were linked up in a head-to-tail manner (visualize an upside-down J fitting inside one that is upright). The next molecule that Rebek created, called ARBI, replicated itself even more efficiently.

ARNI and ARBI really didn't amount to much as life forms. All they could do was duplicate themselves. Real organisms can mutate and evolve. Rebek began to wonder if he could create a molecule that would sometimes "make mistakes." These mistakes (or mutations) might cause it to synthesize molecules that might be better replicators. If he could accomplish this, then a simple kind of evolution would have taken place.

So Rebek sought to see if he could synthesize a molecule that would not only catalyze its own formation, but could also cause the formation of a molecule of a similar shape. He soon found two, which he called ZARBI and ZNARBI. They were not very good self-replicators, but they compensated for this by having the ability to make mistakes. ZARBI would sometimes catalyze its own formation and sometimes make a ZNARBI instead. ZNARBI returned the compliment, sometimes acting as a template for the formation of ZARBI. When the two were placed in competition for reproductive resources, ZNARBI proved to be a slightly better replicator. It would assemble copies of itself from the molecules that it found in solution more rapidly than its competitor ZARBI did.

The next step was to see if these two molecules could mutate. A solution containing ZNARBI molecules was irradiated with ultraviolet light. This removed the "ZN" section of some of the ZNARBI molecules, converting them into our old friend ARBI. ZNARBI now had to compete with ARBI for the chemicals it needed to replicate. The sleeker ARBI molecule, which could reproduce itself more rapidly, won outright. It soon took over the resources of its environment.

Rebek had observed evolution and competition in the laboratory. If he began with ZARBI molecules, then chemical "mistakes" would sometimes produce ZNARBI, which was a better replicator. Ultraviolet light would cause some of the ZNARBI to mutate into ARBI, which was the best replicator of the three. He could begin with a molecule which reproduced itself somewhat inefficiently, and wind up with one that was very good at the job.

Evolutionary change depends, not only on mutation, but also on a process called recombination. The DNA strings found within a cell are arranged in chromosomes. Human cells, for example, contain forty-six chromosomes, which are arranged in twenty-three pairs. They are paired because we inherit one set of chromosomes from each parent. Two paired chromosomes can exchange strings of DNA with one another, creating new genetic combinations. This is advantageous because it creates more genetic diversity for natural selection to work on, enhancing the evolutionary possibilities. (For more on natural selection, see Chapter 3.)

Rebek and his coworkers found that when a self-replicating molecule they called DIXT (diaminotriazine xanthene thymine) was placed in solution with such molecules as ARBI, new combinations arose. One of these, called ART, was the best replicator Rebek had encountered. Another, called DIXBI, was unable to reproduce at all. It was "sterile."

Rebek had shown that synthetic molecules could make copies of themselves, using chemicals they found around them, and also mutate and evolve. But of course these molecules were not yet "alive." Real cells are enclosed in protective membranes that distinguish them from their environment, and they are also able to utilize energy taken from the environment. Animal cells, for example, make use of food the animal ingests, while plant cells depend upon light energy.

So Rebek tried to see if he could create enclosures for his molecules. The first he devised, which he called a "tennis ball," was not really large enough to do the job. It could enclose very small molecules, such as those of methane, but not the replicators that he had created. The next, called the "softball," was much superior. It could hold two of Rebek's self-replicating molecules, which took up about half the space in at the interior.

Softballs turned out to be self-replicating molecules, too. They could split into two pieces, and each half could function as a template from which the other half was formed. Softballs have a preference for enclosing certain guest molecules—those of the right size and shape to fit into them comfortably. Thus the softballs form a kind of symbiotic relationship with other self-replicating chemicals. This is only the beginning. Rebek informed me that twelve of the twenty-five people in his laboratory at the Scripps Research Institute were seeking ways to make certain molecules form enclosures for others.

Work on this encapsulization process is very far removed from research on the origin of life. The encapsulating softballs and other chemicals bear a greater resemblance to the protein coat that protects the DNA or RNA in a virus than they do to a cell membrane. Rebek says it would be easy to create tiny vesicles that were similar to the membranes that enclose cells. But doing so would not aid his research. Real cells contain thousands of different kinds of chemicals. Some of these are present in great numbers. As of this writing, Rebek was working with just a few chemicals at a time. Thus creating small enclosures which would function as reaction chambers was more appropriate. If his molecules were floating around in enclosures the size of cells, they would behave just as they would if they were not encapsulated at all.

What is the next step? Rebek says that he is seeking to create a synthetic cell and to discover the minimum number of enzymes that would allow it to function. If he accomplishes this, there will still be one thing that his synthetic cells will lack. If they are to resemble real cells, they must be able to make use of energy obtained from their environment. At the moment, it is not easy to see how it might be possible to make this happen. Rebek says it is simple to produce chemicals that converted light into chemical energy, but it is "pretty daunting" to integrate them into a self-replicating system. In his 1994 article in *Scientific American*, Rebek said this problem was one for the "next decade." He now believes that this estimate may be a little too optimistic. It may take considerably longer to give his synthetic cells the last characteristic they need to be considered "alive." Whether Rebek is able to create synthetic "living" cells or not, he is likely to discover a great deal about the principles that govern the replication of, and interactions between, organic molecules. He seems to already have made significant steps toward this goal.

Two Outlooks on the Origin of Life

Is there any relationship between Stuart Kauffman's theory of the origin of life and Rebek's work? Probably not. Rebek is working from the "bottom up," trying to fashion minimal synthetic cells. Kauffman sees life as having begun in very complex systems containing large numbers of different kinds of molecules. I thought it might be interesting, therefore, to query Rebek about his opinion of Kauffman's work. The two scientists are doing quite different research, but there are similarities. Both, after all, are involved with the creation of life. Naturally they know one another. Kauffman and Rebek have even filed a patent application together. The application deals with a kind of "random chemistry" that might lead to the development of new drugs.

Rebek made no arguments for or against Kauffman's theory. He did emphasize that it would be very difficult to test experimentally *at present* (his emphasis). He commented that it would be very easy to create large systems of molecules similar to those which Kauffman envisioned. A system of a hundred thousand different molecules, he said, could be made in a few days. Rebek knew of no way scientists could observe minute changes that would be taking place within such a system. And if these small changes couldn't be seen, there was no way of knowing whether the changes going on corresponded to the processes described in Kauffman's theory.

As we saw, Rebek has created two molecules which catalyze one another's formation. This process bears little similarity to the autocatalytic sets envisioned by Kauffman. The latter would contain many thousands of catalysts, not just two. In this case, both theory and experiment have a long way to go before they are likely to meet.

Self-Replicating Proteins

One reason the "RNA first" theory of the origin of life became popular was because it was known that certain kinds of RNA could catalyze the production of other kinds of RNA, while it had never been shown experimentally that proteins could catalyze the replication of anything. Proteins perform many functions within a cell. You may recall that one of them is the splitting apart of strands of DNA, which allows the DNA to reproduce itself, section by section. However, proteins are not directly involved in the replication process.

Then, in 1996, a surprising new experimental result was obtained. Working in the laboratory of M. Reza Ghadiri at the Scripps Research Institute, David Lee and his colleagues created a helical peptide (recall that a peptide is a protein fragment) that was able to replicate itself. The peptide, a chain of some thirty-two amino acids, turned out to be able to assemble copies of itself from fifteen- and seventeen-amino-acid protein fragments, using itself as a template. When the peptide was placed in solution, the reaction initially proceeded slowly. As the number of self-replicating peptides increased, reproduction took place at a faster and faster rate. Then it slowed down again as the supply of the two smaller fragments became exhausted.

One might think that this result would bolster the idea that proteins evolved before RNA. When I asked Ghadiri if he thought that proteins came first, he didn't seem very interested in the question. "Who knows?" he answered, "Who cares?" Ghadiri explains that he is more interested in learning to understand the nature of "life itself." He said there was probably a kernel of truth in a lot of different theories about life's origins, but added that there is no fossil record of the origin of life and of course we can't go back to see how it happened.

Ghadiri thinks that there are three closely interconnected questions about life: 1) What is the nature of life? 2) How did life arise? 3) How can we create life? The second question sounds like a contradiction of some of the statements quoted in the previous paragraph. In reality, it isn't. Ghadiri's position seems to be that we haven't gained an understanding of the origin of life by theorizing about it. Still, an experimental approach might tell us a great deal. If scientists learn to create life, they will certainly have gained some insights into the ways that nature might have gone about the task.

Ghadiri sees life as a kind of biochemical "ecosystem." He thinks that life is the result of dynamic interactions between populations of molecules. When the populations self-organize and produce emergent properties that are greater than the sum of their components, the transition into life begins. He compares a living organism to an ant colony. An individual ant has a simple nervous system and does not behave in very complex ways. Ants interact with one another using only about a dozen different signals. Ant signaling is at a level far below human conversation. Yet an ant colony is a kind of superorganism; collectively, the ants are capable of very complex kinds of behavior.

Ghadiri wants to find out how the transition from the molecular world of simple chemical reactions to a living "ecosystem" happens. To this end, he has devised experiments in which sets of molecules are likely to exhibit emergent properties. In one of them, a self-replicating peptide was placed in a solution with smaller pieces that were mutated versions of parts of the original peptide. This caused the original peptide to assemble mutant copies of itself.

The experiment worked something like this. Ghadiri's self-replicating peptide, called T (no relation to the nucleotide T), could make copies of itself from the smaller molecules E and N. E and N would line up on T as a template, making a new T. The second T could then drift away and repeat the process when it encountered another E and another N. When mutant forms of E and N were added to the solution, three types of mutant T were created. One of these was made up of a normal E and a mutant N. In the second, it was E that was of the mutant variety. In the third, both the E and N fragments were mutants.

You might expect this would create a situation where the four different versions of T competed for resources and the one that replicated most efficiently would "win." The original peptide did win out, but not because it was the most efficient replicator. The mutant peptides containing one particular mutation didn't make copies of themselves at all; they created the original unmutated T instead. Meanwhile, the T that contained both a mutant E and a mutant N just didn't replicate. A dynamic error-correction mechanism had spontaneously developed.

When reading this account, it is easy to get bogged down in all the Ts and Es and Ns, and lose sight of what is going on. To summarize the results, Ghadiri had developed a mechanism that prevented mutations from proliferating too rapidly. This is important because if there are too many mutations in a biological system, its genetic code rapidly turns into garbage. There must be some mutation if evolution is to take place. But if the mutation rate is too high, the system will simply self-destruct.

One reason DNA is double-stranded is that if an error is made, it can be fixed; the correct code will still be present in one of the strands. It is perfectly accurate, by the way, to speak of "reasons."

After all, DNA didn't evolve by chance. It evolved because it was superior to single-stranded RNA as a repository of genetic information.

Ghadiri's result was important because it showed that a mechanism that guarded against an excessive accumulation of errors could spontaneously arise. Furthermore, it was not necessary to create DNA to create such a mechanism. It appeared as an emergent property in solutions containing small pieces of protein.

In yet another experiment, Ghadiri placed three peptide pieces in solution. They were able to combine with one another to form two different replicating peptides. Again, you might think that a "survival of the fittest" kind of situation might result, and that the peptide that was the better replicator would win out. This isn't what happened. The two peptides became symbiotic. A situation developed where each catalyzed the formation of the other one. Furthermore, each catalyzed the formation of the other more efficiently than it replicated itself. In other words, A wasn't very good at making more As, but it did a pretty good job of creating Bs. The behavior of B was similar. It was better at making As than replicating itself.

These are only the first steps in Ghadiri's quest to understand the nature of "life itself." He plans to assemble systems that are increasingly complex to see what kinds of emergent properties arise. He also wants to learn something about the ways in which a chemical system will adapt to its environment. He thinks that if a system of chemicals begins to receive different stimuli from the outside "world," it might be able to reorganize or "rewire" itself and exhibit new kinds of behavior. Here, he makes an analogy with the self-organization of human populations. He points out that external "stimuli" can cause these populations to self-organize themselves in numerous different ways. If there is a disaster, such as an earthquake, a human population will reorganize and exhibit new kinds of behavior. We are currently seeing this in California, where a great deal of emphasis is being placed on earthquake retrofitting of buildings, bridges, and other structures. This is being done because both northern and southern California had large earthquakes in recent years.

There is reason to think that something similar might happen in systems of biochemical molecules. No one knows how they might reorganize themselves, or what the effects might be. All we know is that there seems to be a reasonably good chance of discovering new types of phenomena.

Ghadiri talks about the emergent behavior of sets of molecules. So does Stuart Kauffman in his theory of autocatalytic sets. Ghadiri created something that bore a distant resemblance to an autocatalytic set when he found two peptides that catalyzed one another's formation. This might cause you to wonder how deep the connections are between Kauffman's theory and Ghadiri's experimental work. When I asked Ghadiri about this, he replied that he believed chemical systems were capable of self-organizing in many different ways, and that the creation of an autocatalytic set was just one. He said that he thought systems would turn out to have the ability to organize themselves in ways we haven't even begun to imagine. He then went back to talking about the transition from the molecular world to living ecosystems and about his quest to discover the nature of "life itself."

3

THE EVOLUTION
OF COMPLEXITY

In his 1996 book *Full House*, Harvard paleontologist Stephen Jay Gould argues that there is no evolutionary trend toward increasing complexity. Bacteria, he says, are and always have been the dominant form of life on earth. He points out that, during the course of an individual's life, the number of *E. coli* bacteria (*Escherichia coli* help us digest our food; they are usually beneficial, though certain strains can cause disease) in the intestines of a human being far exceeds the number of people who have ever lived. Furthermore, bacteria may contribute more to the total biomass of the earth than all other living organisms combined.

Natural selection, Gould says, simply causes organisms to be adapted to their environments. There is no trend toward "progress." A mammal is not really a "higher" form than a fish, and human beings do not represent the apex of evolution. Evolution has no goals; it simply weeds out organisms that are not well adapted to their environments.

Gould does not deny that complex life forms exist. He views the emergence of complexity as the result of random evolutionary events. If life requires some minimum level of complexity, he points out, randomness will ensure that a small number of organisms will be more complex than the average. Gould further emphasizes that many organisms are simpler than their predecessors. Parasites, for example, are simpler than their free-living ancestors. If an organism can attach itself to a host, it doesn't need all of the anatomical features that its evolutionary ancestors possessed. There is evidence that there are more species of parasites than there are of other organisms.

Virtually all biologists would agree with the idea that equating evolution with progress is a fallacy. Darwin himself repudiated the idea. The fallacy lingers on for a number of reasons. Our inclination to view human beings as "higher" on the evolutionary scale than other animals certainly plays a role. Also, we must remember that Darwin's theory of natural selection was a product of the Victorian Age in England. Belief in progress was practically universal in those days, and the idea that science could cure all of humanity's ills was widely held. It was an age in which scientific tomes could become bestsellers, and an age in which the working class would flock to hear famous scientists lecture. It seemed only natural to think that, if progress was characteristic of human society, it could be seen in the biological world also. It may seem strange that the idea has lingered on for so long, but it apparently has. On the other hand, the equation of evolution with progress may have speeded the acceptance of Darwin's theory. Though there was some opposition from the creationists, just as there is now, acceptance of the theory was fairly rapid.

But there is no progress in evolution. Evolution knows nothing of goals. On the contrary, it is based on the workings of chance. Natural selection operates on chance mutations, preserving those that are beneficial and eliminating those that are deleterious. There is nothing more to it than that. The organisms that inherit a makeup that causes them to be well adapted to their environments survive and reproduce. Those that possess genetic makeups that make them less well adapted do not survive. An organism does not have to be more complex than its predecessors in order to survive.

Gould's claim that there are no trends toward increasing complexity is somewhat more controversial. Not all biologists would agree with him. In his book, *The Blind Watchmaker*, Oxford University zoologist Richard Dawkins takes a somewhat different point of view. He feels that the existence of complex organisms is something that must be explained and calls it the "deepest of problems." According to Dawkins, natural selection builds complexity in small steps. Over periods of many millions of years, evolution has gradually created complex structures in living organisms. In order for this to happen, it is necessary that each small increment enhance a species' chances of survival. Over periods of many millions of years, this can lead to the evolution of such complex structures as the eye or complicated mammal brains. Dawkins says Darwin's theory is the only one known that can solve "the mystery of our existence." Dawkins is, of course, referring to the existence of biological complexity.

One might think that progress and increasing complexity are the same thing. This misinterpretation is one of the factors that has caused belief in progress to persist. Once we give up the idea that more complex organisms are somehow "higher" on the evolutionary scale than are simpler ones, the problem disappears. Human beings are well adapted to their environment (including environments they have created). So are cockroaches. Each has performed the job of adaptation relatively well. One could turn the argument about progress on its head and say that the cockroach is the obvious higher form. After all, it has survived in its present form for about 300 million years, while homonids (members of the genus *homo*, and some of its predecessors) have existed for only 2 million. There is every reason to think that the cockroach will continue to survive long after we are gone.

One must concede that Gould has a point. Our view of life is often somewhat anthropomorphic. We pay the most attention to the birds, fish, and mammals that we see, and sometimes imagine that our interest in them is a measure of their importance. These animals constitute only a small minority of all living organisms. For example, there are only about 4,000 different species of mammals. But there are about 750,000 species of beetles alone. No one knows how many millions of different kinds of bacteria exist.

In any case, we can't be blamed if the more complex organisms—including other human beings—attract our attention the most. After all, if nothing more complicated than bacteria had ever evolved, life on earth would not be very interesting, except perhaps to extraterrestrial visitors seeking evidence that life had evolved in many places in the universe. Thus it is perfectly legitimate to ask how the more complex life forms, such as multicellular plants and animals, came about. Even those who agree most wholeheartedly with Gould cannot deny that biological complexity exists.

Gould is almost certainly right when he says that progress and increasing complexity are not the same thing. He may very well be right when he says that there is no trend toward the more complex. One can create an analogy to his argument. Imagine flipping a coin a number of times. If there are only a few flips, it is unlikely that any meaningful (meaningful to us, not the coin, which is operated on only by chance) patterns will appear. But if the coin is flipped a billion times, it is likely that one will see some very complex patterns. In this case, just as in evolution, the appearance of complexity is not the result of progress toward some goal. Chance alone is sufficient to explain it.

Then why do such notable scientists as Gould and Dawkins disagree? I suspect they are looking at different aspects of evolution. It is true that most organisms on our planet have remained relatively simple. But it is also reasonable to ask how the more complex ones evolved. Perhaps the existence of a few very complex organisms can be attributed to the workings of chance. But exactly what was going on when evolution made them complex? It is perfectly reasonable to attribute the existence of complexity to the workings of chance and go on asking what was going on when it appeared.

Defining Complexity

It is not easy to define biological complexity, however. Is a human being more complex than a lungfish? In at least one sense, we are not. Each cell of a lungfish contains about fifty times more DNA than is found in human cells. Frogs also possess a great deal of DNA. So, judged by this measure, frogs and lungfish are complex, and human beings are relatively simple. It is also possible to arrive at the opposite conclusion. Organisms with a lot of DNA tend to have large cells. Consequently they will have fewer cells in their brains than comparable creatures with smaller amounts of DNA. It's easier to pack a lot of small items into a given amount of space. Similarly, though there are the same number of vertebrae in the neck of a human being and the neck of a giraffe, each has many fewer total vertebrae than a snake. Thus, in one sense, snakes are more complex animals than we are.

Complexity may be hard to define. However, there have been times in evolutionary history when no one can deny that complexity definitely increased. The first was the evolution of eukaryotic cells. Bacteria are prokaryotes (pronounced pro-carry-oats). Their cellular structure is relatively simple. Bacterial cells have no nuclei, for example. Eukaryotes (you-carry-oats), on the other hand, are noticeably more complex. Their DNA is contained within a nucleus, which is enclosed within its own membrane, and there are numerous other structures, called organelles, within the cell. I won't attempt to name them all. For our purposes, it will be sufficient to mention that there are some eleven different organelles in animal cells that are not found in bacteria.

Of particular interest are the mitochondria, which convert energy-rich food molecules into forms of energy that can be used by the cell. Mitochondria are found in both animals and plants. Plants also contain chloroplasts, which carry out photosynthesis. Both kinds of organelles carry their own DNA and reproduce independently of the cells that enclose them. Mitochondria—but not chloroplasts—have a genetic code that is slightly different from that of the cells in which they reside. Their DNA triplets sometimes code for different amino acids than the "normal" DNA in the nucleus of the cell. This is really very surprising. The "normal" DNA code, after all, is the same in all living organisms. DNA triplets code for amino acids the same way in human beings that they do in the simplest bacteria. It is only in mitochondria that coding is sometimes different.

It is impossible to date exactly the steps that led to the evolution of organelles and eukaryotic cells. Bacterial fossils that are billions of years old have been found, but it is often impossible to determine what structures existed inside them. However, it is thought that the predecessors of modern eukaryotic cells began to evolve about 3 billion years ago, and that cells with nuclei existed 2 billion years ago. These primitive cells contained no mitochondria or chloroplasts, however. Finally, around 1.5 billion years ago, cells with mitochondria and chloroplasts evolved, and eventually gave rise to multicellular animals, plants, fungi, modern single-celled animals (such as amoebas), and algae.

It is not known exactly when multicellular forms first appeared. Fossils of organisms that might have been multicellular have been found in 2-billion-year-old rocks, but it is not possible to draw any definite conclusions about them. It is possible to determine, however, that by about one billion years ago, eukaryotic life was abundant and diverse. It owed its success to the acquisition of mitochondria and chloroplasts. Life gets somewhat easier if you possess little energy powerhouses (mitochondria), or organelles (chloroplasts) that will do the job of photosynthesis for you.

It was noticed as early as 1890 that mitochondria might be "organisms" similar to bacteria. In 1910 it was suggested that chloroplasts arose from a symbiotic relationship between photosynthesizing bacteria and primitive nucleated cells. For decades, such ideas were ignored or ridiculed. It wasn't until the early 1960s, when it was discovered that mitochondria and chloroplasts contained their own DNA, that scientists began to seriously consider the idea that the presence of these organelles in eukaryotic cells was due to symbiotic relationships that developed over a billion years ago. The acceptance of the theory is largely due to the work of the American biologist Lynn Margulis.

It is now thought that eukaryotic cells developed when primitive nucleated cells ingested smaller bacteria. Over a period of a billion or more years, a stable relationship developed between the host cells and their symbionts, and the ingested cells lost functions that were no longer needed in their new environment, the host cell. Thus they evolved into the chloroplasts and mitochondria that we observe today. There are a number of kinds of evidence that support this idea. One, the finding that chloroplasts and mitochondria have their own DNA, has already been discussed. Furthermore, nucleotide and protein synthesis in chloroplasts and mitochondria exhibit numerous similarities to these processes in bacterial cells. This is considered to be strong evidence that they evolved independently.

Finally, it is possible to observe similar symbiotic relationships today. For example, certain marine slugs and related animals feed upon green algae. The chloroplasts in the algae continue to carry out photosynthesis for some time after they are incorporated into the slugs' cells, and the carbohydrates produced by photosynthesis become a source of nutrients for the slugs. The chloroplasts do not grow or divide, but the relationship can continue for periods of months. The fact that chloroplasts can be taken up by animal cells and continue to function gives support to the idea that they formed symbiotic relationships in the past. If they can become symbionts of animals today, this is evidence that they are especially adapted to this role.

As I remarked previously, there are many types of organelles besides mitochondria and chloroplasts that are present in eukaryotic cells, but I have said little about them. The reason is that much less is known about their origin and evolution. They do not contain their own DNA, so one cannot say whether they owe their origin to ingested bacteria which became symbionts, or whether they are structures that evolved within nucleated cells. However, the evidence that mitochondria and chloroplasts evolved from symbionts is dramatic.

Multicellular life—at least the multicellular life from which modern plants and animals are descended—first evolved about 600 million years ago. It is not difficult to imagine how this happened. Single-celled organisms must have formed colonies. We encounter colonies of single-celled organisms today. Corals are just one of many examples.

Fossil evidence indicates that even the more primitive prokaryotic bacteria sometimes formed colonies. As the cells that were members of the colonies evolved, some of them took on specialized functions, and multicellular plants, animals, and fungi eventually developed.

Today, there exist "animals" that are intermediate in form between single-celled organisms and true multicellular life. An example is the jellyfish known as the Portuguese man of war. This jellyfish is a colony of hundreds of different individuals that are genetically identical. However, these individuals assume many different shapes and functions. One organism shapes itself into a float, which keeps the colony on the surface of the water. Other individuals become specialized for capturing prey and digesting food, or become male and female reproductive organs. Although the tentacles look very different from the float to the human eye, they share the same DNA, just as the cells in a human body do. The only real difference is that the Portuguese man of war is "multi-individual," where we are multi-cellular. The Portuguese man of war has no brain or system of neurons that sends messages to the cells. There is no particular reason to believe that modern plants and animals evolved from colonies like the Portuguese man of war; it is possible that this jellyfish may simply represent a different path in evolution. However, the existence of such creatures makes the evolution of multicellular life seem somehow inevitable. If multi-individual colonies can evolve, the evolution of true multi-cellular life should not be particularly difficult.

The Cambrian Explosion

The evolution of multicellular organisms was not something that happened quickly. As many as a billion years may have passed between the evolution of the first eukaryotic cells and the appearance of the first multicellular organisms. The evolution of the latter might have been impeded by a lack of oxygen in the earth's atmosphere. Recall that there originally was little or no free oxygen on the earth. Levels were built up gradually by the action of photosynthetic bacteria over a period of billions of years.

A substance known as collagen is the primary connective material in animals. But collagen cannot be manufactured without oxygen. Oxygen levels approaching those of the present day earth are also required if structures such as gills and circulatory systems are to develop. You should note that I emphasize gills. Initially, multicellular life existed only in the oceans; it did not migrate onto land until many millions of years later.

Multicellular life first evolved about 600 million years ago. The fossil record for this period is rather sparse, and this has hampered the investigations of scientists who have attempted to study it. In some cases, the fossils consist of little more than markings left in sediment by what appear to have been burrowing worms. And then, about 530 or 535 million years ago, something new happened. Numerous strange and wonderful forms suddenly appeared. This happened so rapidly (rapidly, that is, on the geological time scale, where an "instant" can really be millions of years) that biologists often speak of the "Cambrian explosion." It is called "Cambrian" because it occurred during the geological period of the same name. The Cambrian period lasted from around 530 million (or 535 million) to around 505 million years ago.

The Burgess Shale

The Burgess Shale is a small limestone quarry, formed 530 million years ago, located in the Canadian Rockies. In 1909 the American paleontologist Charles Doolittle Walcott discovered that the quarry contained fossils of numerous, previously unknown marine animals. Although many of them looked very strange, Walcott erroneously classified every one of them as a member of a modern group, viewing them as ancestors of creatures alive today. Thus he delayed a scientific discovery that was made decades later. That Walcott classified the fossils the way he did was not entirely due to his being a victim of his own preconceptions. He may never have found the time to study them properly. His administrative duties as secretary (i.e., director) of the Smithsonian Institution were apparently quite burdensome. Meanwhile, other scientists refrained from studying them because they did not want to usurp Walcott's right to study and describe them. Even today, paleontologists hesitate from infringing on one another's turf.

It wasn't until the 1970s, when scientists, notably Harry Whittington of Cambridge University, began to reexamine the specimens, and it was realized that many of them bore little resemblance to anything living today. The Burgess Shale specimens included many "weird wonders" (a term Stephen Jay Gould uses to describe them in his book, *Wonderful Life*). One, for example, was an animal that had five eyes and a frontal nozzle that somewhat resembled the hose of a vacuum cleaner. Another was a stalked animal that looked something like a flower, and had a mouth adjacent to its anus. Others were almost as strange. Of course some did appear to be ancestors of living creatures. But the disparity of the life that existed during the period when the fossils were formed was amazing.

Fossils resembling those in the Burgess Shale have since been found at other geological sites. But not at a great number of them. Many of the fossilized creatures from this era were soft bodied. And soft-bodied animals do not fossilize very well. The soft parts of an animal generally decay. This is why fossils that are found generally consist of bones and other hard parts. It is estimated that only 20 percent of the animals living in today's oceans would fossilize easily, for example. Soft-bodied animals are likely to be preserved only in environments lacking oxygen. But environments without oxygen are not places where animals who need it can live. It is thought that the Burgess Shale animals were buried by a mudslide that carried them down to a location in deep water where little of the gas was present. This mudslide may have taken them to a quick death, but it was to become a boon to science.

At this point, I should probably digress a little and introduce some biological jargon. Paleontologists sometimes distinguish between biological "diversity," or the existence of many different species, and "disparity," which means a large variety of different anatomical designs or body plans. All mammals conform to the same body plan. For example, the bones in the forelimbs of a human being can be matched up with the bones in the wings of a bat or the flippers of a whale. On the other hand, there is no such correspondence between parts of mammal anatomy and the anatomy of an insect.

When organisms differ greatly in their body plans, they are said to belong to different phyla. Today there are about thirty animal phyla. One phylum is that of the arthropods, which includes insects, spiders, and crabs. Another is that of the mollusks, which includes snails, clams, and squids. Yet another phylum is that of the chordates, to which human beings (and also such organisms as sea squirts) belong. The largest subgroup of the chordate phylum is that of the vertebrates, which includes birds, fishes, and mammals.

When I say that there are "about" thirty phyla, this is a reflection of the disagreement between taxonomists as to how animals should be classified. Some will place two groups of organisms in the same phylum, while others will create two separate classifications. Something like this happens even at the species level, even if the species in question is well represented in the fossil record. Until the late 1990s, paleontologists were unable to agree as to whether the Neanderthals were a subspecies of *Homo sapiens* that was adapted to glacial climates, or a separate species. Even though Neanderthals were plentiful in the fossil record, and even though a great deal was known about their culture, the controversies continued. It was only in 1997, when some DNA was extracted from Neanderthal bones, that the question was finally decided (yes, they were a separate species). Of course, there are some distinctions that scientists (and nonscientists as well) find it easy to agree upon. No one would argue, for example, that dogs should be placed in the same phylum as clams.

The word phylum is just a technical term for body plan. It is not the highest level of classification. There is a higher level, that of "kingdom." There are five kingdoms. Multicellular organisms are divided into plants, animals, and fungi, while single-celled organisms are either protists (those with complex eukaryotic cells) or the prokaryotic organisms that we generally refer to as "bacteria." Thus a phylum is a major division of animal life.

The most striking thing about the Burgess Shale fossils (and fossils that have been found at other sites) is that many of them do not seem to fit into modern phyla. There is no general agreement as to how many phyla there were then which do not exist today. Gould, for example, is of the opinion that fifteen or twenty of the organisms found in the Burgess Shale are so different from anything now living, that a new phylum should be created for each. Other scientists express the more conservative opinion that many of these strange animals would be placed in existing phyla if we better understood their structures. However, it cannot be denied that the disparity of life was much greater in the Cambrian period than it is now. Today, many more different species exist. Then, there was much greater variety in body shape and design.

The disparity was greater within phyla too. Today there are three major kinds of arthropods: crustaceans (a group which includes crabs and shrimp), chelicerates (spiders and scorpions are members of this group), and uniramians (mostly insects).

The Permian Extinction

The earth has experienced a number of mass extinctions. The most famous occurred 65 million years ago, in which the dinosaurs perished. It is now generally agreed that this extinction was caused by the collision of an asteroid with the earth. The collision left an enormous crater about 180 miles in diameter beneath the Yucatan peninsula of Mexico. This event caused over half of all living species to become extinct, and allowed mammals to proliferate in a world that had previously been dominated by dinosaurs. Before the extinction, mammals had been small animals no larger than modern rodents. Once the dinosaurs were gone, they expanded into ecological niches that were now unoccupied.

The mass extinction that wiped out the dinosaurs is not the largest the earth has experienced. In the Permian extinction, which took place about 245 million years ago, not half, but ninety-five or ninety-six percent of the living species perished (the Permian is another geological period; it lasted from about 285 million years to about 245 million years before the present). There is no evidence, by the way, that this extinction was caused by a collision with an extraterrestrial body. Collisions seem to have caused some extinctions, but not all of them. There have been times when there have been major impacts that have not produced any significant number of extinctions. For example, two large bodies collided with the earth around 36 million years ago. But the next major extinction event happened some two million years after that.

Many scientists link the Permian extinction to the coalescence of the earth's continents into a single supercontinent, called Pangaea. The continents can be thought of as gigantic floating rafts. Continental crust is made of granite, which is lighter than the basaltic rock that lies beneath it. Continents can either drift apart or coalesce. The drift is not very rapid. It amounts to no more than a few centimeters per year. Over long periods of geological time, the effects can be dramatic. For example, mountain ranges are created when continents collide with one another. When they drift apart, they sometimes leave telltale marks. Anyone who has ever compared a map of South America with one of Africa cannot escape having the feeling that the two continents were once joined. South America bulges outward where Africa curves inward. And yet today, a large stretch of the Atlantic Ocean separates the two continents.

The coalescence of the continents into Pangaea may not have been the only cause of the Permian extinction, but it certainly must have contributed to it. At this time, most organisms lived in marine environments, usually in shallow water. After the continents came together, there was much less coastline, and thus many fewer places where marine organisms could live. Of course there must have been other significant environmental changes as well. Climatic conditions would certainly have changed, for example.

We should not think of the species that survived as having been somehow "fitter" than those that died out. Evolution adapts species to their environments, and if the environment changes, survival may be a matter of luck. There may be chance elements in their anatomical designs that fit them to their new environment. Organisms with the ability to become dormant when environmental conditions become unfavorable are more likely to survive an extinction even though they may not derive great advantages from this ability in more normal times. Sometimes it is difficult to see why some organisms survive while others do not. Clams were relatively unaffected by the Permian extinction, while other molluscs died out. It doesn't appear that there is anything very magical about being a clam, yet the clams turned out to be survivors.

I don't want to dwell on the subject of surviving an extinction. That would take me too far from my real subject, that after the Permian extinction, life diversified in a very different way than it did during the Cambrian. New species evolved. And if we use the number of different species as a measure of the complexity of the biosphere, we can say that, after the catastrophe, the biosphere again began to grow more complex. But somehow it was not the same.

Disparity versus Diversity

You might think that if 95 percent or more of all living species were wiped out in the Permian extinction, this would have set the stage for another great period of evolutionary experimentation like the one that took place in the Cambrian. This did not happen. It is true that many new animal species evolved after the extinction. Yet no new phyla appeared. Existing body plans were altered, but no great innovations took place. In other words, there was an increase in diversity, but not in disparity.

Here I should point out that when life first began to move onto land about 400 million years ago, new forms did develop. In the case of plants we speak not of phyla, but of divisions. Seven new divisions of plants evolved at this time. Nothing of the sort happened in the case of animals. No new animal phyla were created, even though life was moving into previously unexplored ecological niches.

It is thought that the Cambrian explosion lasted five to ten million years. Thus the range of anatomical variety reached a maximum shortly after multicellular organisms evolved. After that, evolution worked on designs that already existed. Consequently, there is less disparity today than there was 530 million years ago.

There are several theories which seek to explain why this should be the case. According to the most popular one, the Cambrian explosion took place at a time when there were many available empty ecological niches. There were ample food supplies, and the animals that were evolving at that time initially had little competition. Thus almost any anatomical design would work. Numerous phyla evolved because life was evolving into an empty environmental space. After the Permian extinction, members of surviving phyla filled any empty niches more effectively than any newly-evolved (and thus less well adapted) new phyla could have. Hence there was less wholesale experimentation and more fine-tuning.

According to a second, more recent, theory, the cause had more to do with genetic stability. In order to understand this idea, it might be a good idea to review and expand upon what has been previously said about the ways in which genes interact with one another. It is usually not the case that a single gene will code for a specific trait in an organism. Genes can act on other genes, turning one another "on" or "off." There are also gene complexes that work together. For example, five different genes may produce five different protein enzymes that synthesize a sixth chemical that is not produced by any gene acting alone. The sixth chemical may play an important role within the metabolism of the cell, and be related to certain biological traits. Finally, there are some genes which produce proteins which affect the mutation rate of other genes. And of course, any individual gene may be a member of more than one network. The interaction of genes is vital to the existence of complex organisms. For example, every cell in a human body contains the same genetic code. If there were no interactions between genes, each cell would be identical. There would be nothing that could cause differentiation between liver cells and skin cells. There would be no way to create specialized red blood cells, the cells that form the lining of the stomach, or even the neurons of the brain.

It is possible that there might have been fewer interactions between the genes of the Cambrian organisms than there were in later ones. In some sense, genetic systems may have later "aged" or "matured." If this was the case, the organisms that existed after the Permian extinction would be expected to have less genetic flexibility. Genetic patterns could have been "locked in" in such a way as to prevent any large departures from existing body plans. Naturally, this is pure hypothesis. We can't go back to the Cambrian period to determine what genetic systems of that era were like. Fossil remains tell us only what the animals looked like; they convey no information about their internal chemistry. But there is probably some grain of truth in this idea. Naturally it doesn't contradict the first, more popular, theory. Both factors could have been operative.

A third theory is one that has been proposed by Stuart Kauffman. It is based on computer modeling of genetic systems. But before I cover it, I must explain some bits of jargon that are commonly used by evolutionary biologists these days.

Biologists often speak of "fitness landscapes," which are visualized as having numerous peaks and valleys. The peaks correspond to high fitness, while a species that occupies a valley is one that is poorly adapted to its environment. Natural selection will thus cause species to ascend toward the peaks for the simple reason that organisms inhabiting the valleys quickly die out. The higher the species climb, the better their adaptation. Consequently, once a new body plan evolved, natural selection would alter it in such a way to bring it near a peak.

It is not necessary to go into all the mathematical details of Kauffman's theory to grasp the basic idea. During the Permian extinction, numerous species were wiped out. However, most Cambrian body plans survived. According to Kauffman this may have been because most or all of the adaptive peaks were still occupied. If any new phyla were to be created, very long jumps (corresponding to large mutational changes) to distant, unoccupied, peaks would have been necessary. But, according to Kauffman's mathematical model, such long jumps are likely to be unsuccessful; it is far too easy to land in a valley. Kauffman calculates that the chance of success decreases rapidly with the length of the jump. In other words, species stay pretty much as they are, with no dramatic changes taking place. This idea isn't difficult to understand. If you were climbing in the Alps, and if you possessed some kind of magical ability that allowed you to jump from one place to another, a short leap would probably not produce a large change in altitude. Yet a very large jump might very well land you in a deep valley, or even in some lowlands that were not even part of the mountain chain. In real life, this might not be any great disaster. In biological evolution, inhabiting a place at too low an altitude leads to extinction.

Kauffman's theory was by no means the last word on the subject. In 1997, another hypothesis was advanced. A team of geophysicists, headed by Joseph L. Kirschvink of the California Institute of Technology, announced they had discovered evidence that, around the beginning of the Cambrian explosion, continental drift suddenly began to proceed at a much greater rate than it has in other geological periods. During a geologically short period of about 10 or 15 billion years, the continent that later became North America moved from a position near the South Pole all the way to the equator. Meanwhile, the other continents, which were then grouped together in a supercontinent known as Gondwana, traveled across the Southern Hemisphere.

Kirschvink and his colleagues attribute this rapid movement to shifting masses in the earth's interior which caused our spinning planet to become imbalanced. They say that the entire surface of the earth reoriented itself to compensate. The evidence supporting the hypothesis comes from studies of the magnetic orientations of ancient rocks. When rocks form, the magnetic materials that they contain will align themselves with the earth's magnetic field. This makes it possible to determine the location of the rock at the time that it was created. A rock near the equator when it was formed will have a magnetic orientation very different from one that was near a pole.

If the continents were rapidly moving during the beginning of the Cambrian period, this would have caused rapid changes in environmental conditions. This could well have contributed to the rapid growth in the disparity of life, since evolutionary pressures would have been greater than they have been at any time before or since.

As I write this, the hypothesis must be considered a very tentative one. Kirschvink and his colleagues have made large claims that are based on relatively small quantities of data. If further research confirms their findings, they will have made a discovery of tremendous significance. W. D. Dalziel, a University of Texas geophysicist, probably summed up the situation best when he said, "If it's true, it's incredibly important. But it is based on a somewhat scarce database that is open to many different interpretations."

At this point, you may ask what conclusions can be drawn about the reasons why the Cambrian explosion took place. There may be some truth in all of the theories that I have discussed. There are similarities between the first theory (about empty ecological niches) and Kauffman's mathematical model. Speaking of filled ecological niches and occupied adaptive peaks are really not such different things. Evolution adapts species to their environment (the ecological niches that they find) and thus nudges them toward the peaks in the fitness landscape. The major way in which Kauffman's theory differs from the standard view is that it is based on computer models, which he says have demonstrated the improbability of long jumps.

Kauffman, incidentally, thinks that there may be parallels between biological and technological evolution. He points out that after the bicycle was invented, numerous different designs were developed. There were models with big front wheels, and ones with big back wheels, and bicycles with more than two wheels in a line. But today, most bicycles are refinements on one basic design. I'm not sure that the analogy is a perfect one. There may have been a lot of different bicycle "phyla" at first, just as there were a lot of biological phyla during the Cambrian period. But there aren't a lot of different bicycle "species" today. As Kauffman says, nowadays we only have street, racing, and mountain bikes. If Kauffman is right, then it is conceivable that evolution may tend to work in certain characteristic ways, whether it is biological or technological evolution. Of course we shouldn't place too much faith on this analogy. Analogies are notorious for being misleading.

It has always been assumed that the disparity that developed during the Cambrian explosion could be attributed solely to natural selection. It now seems that this may not be the case. The plausibility of some of the theories I have described, Kauffman's in particular, indicates that the increase in disparity during the Cambrian period may be only partly attributable to natural selection. It is entirely possible that an explosion in the disparity of species under certain conditions is an emergent property. It is not only cells and organisms and gene interactions that are complex. An evolutionary or ecological system is highly complex also.

When I say this, I am only surmising that it *might* be true. The Cambrian explosion may have taken place because life was simpler 530 million years ago than it is now. The genetic complexity of multicellular organisms has been evolving for hundreds of millions of years. If the genes of Cambrian organisms operated in simpler networks than is typical of genetic systems today, the rapid environmental changes with which they had to contend may well explain why so many strange new life forms suddenly appeared. It might very well be easier to change a body plan than it is to develop complexity in genetic systems.

Perhaps it would be best not to engage in too much speculation on this point. It may be that we will never know what caused a sudden increase in life's complexity during the Cambrian explosion. In this respect, the problem is similar to that of the origin of life. We know that life evolved at least 3.5 billion years ago, but we cannot go back in time to see how it happened. Similarly, we cannot go back and perform experiments that would tell us exactly what it was that caused a great increase in the disparity of life during the Cambrian period.

Can the Sciences of Complexity Study the Emergence of Complexity?

The term "sciences of complexity" is used to describe investigations of the emergent properties of complex systems. What about the emergence of complexity itself? It appears that, almost as soon as we began to consider problems related to the evolution of species, this question was immediately raised. Evolution is itself a complex system. Unlike many others, it is a system that gives rise to increasing complexity as an emergent property. Whether or not there is any real trend toward complexity in the short term, we can't deny that some of the organisms existing today are much more complex than those that lived 3.5 billion, 2 billion, or even 600 million years ago. Is it really possible to say something meaningful about why this happened?

For that matter, it is possible to go at least a level further. Mammals exhibit greater complexity than bacteria. Animal societies evolve. The ant must be descended from some solitary creature. Yet today there are ant colonies. And evolution has undoubtedly caused them to become more complex over time.

Human societies have evolved also. They have done so in myriad different ways. If they had not, this would be very surprising. Human beings, after all, are quite complex compared to such social animals as ants and bees. It is easy to see that we already have three different levels: the complex system we call evolution, the complexity that evolution gives rise to, and increasing complexity in the social interactions that exist once complex organisms evolve.

As we shall see in subsequent chapters, the study of complexity within complexity, and complexity at even higher levels, can indeed be addressed. If it could not be, then there would be little hope that the sciences of complexity could tell us very much about the evolutionary process. The origin of life and the nature of genetic networks are interesting topics. But life did not remain as it was once it came into existence. And there is more to life than a set of genes which pass on heritable characteristics from generation to generation, and which interact with one another in complicated ways. If we are going to gain any insights into the possible answers to the question, "What is life?" we must find new ways of understanding life's history.

4

FITNESS LANDSCAPES

The concept of evolution did not originate with Charles Darwin. In fact, the idea that species might evolve was often discussed in Victorian England long before Darwin published his book *On the Origin of Species* in 1859. Darwin's grandfather, the physician Erasmus Darwin, had been an evolutionist. In his 1794 book, entitled *Zoomania*, Dr. Darwin had proposed that "irritations, sensations, volitions, and associations" caused animals to adapt themselves to their environments; and their offspring inherited the resulting changes. Dr. Darwin believed the earth was millions of years old, and speculated that all life might have descended from a single organism.

Even in Erasmus Darwin's day, the concept of evolution was an old one. Evolutionary ideas can be found in ancient Indian and Chinese thought and in the teachings of the pre-Socratic Greek philosophers. It is true that speculation about evolution didn't continue for long in the Western world. Aristotle was opposed to the idea, and his authority put a stop to speculation on the subject. However, the idea of evolution was apparently such a natural one that scientists were often talking about it by the mid-eighteenth century. Indeed, it was around the middle of the century when the French naturalist the Comte de Buffon suggested that apes and human beings might have a common ancestor. He didn't place much emphasis on the idea, however; it was just one of many speculations about evolution that were scattered throughout his forty-four volume work, *Natural History*.

It was easy to suggest that species might evolve into other species. Explaining what could cause evolution to take place was a much more difficult problem. The first consistent theory of evolution was proposed in 1809 by the French naturalist and philosopher Jean Baptiste de Lamarck. He is the scientist who coined the word "biology." Though Lamarck's theory turned out to be wrong, it is worth saying a few words about it. When I later contrast Darwin's theory with Lamarck's, it will be easier to understand the magnitude of Darwin's achievement.

According to Lamarck, all living organisms had a built-in drive toward perfection that caused them to rise on the evolutionary scale. There was a kind of inward striving toward greater complexity. "Excitations" and "subtle and ever-moving fluids" caused the "lower" forms to continuously transform themselves into "higher" ones. The belief that there existed "lower" and "higher" organisms was already common in those days. The idea can be dated all the way back to Aristotle. This concept of a "Great Chain of Being" should not be confused with the idea of evolutionary progress. That some beings were naturally "higher" than others was thought to be part of the natural order of things.

Evolutionists viewed the matter somewhat differently. Lamarck believed there was an upward striving toward higher rungs on the Chain, and this striving showed itself in very specific ways. Organisms, after all, needed to adapt themselves to their environments. Lamarck used the recently discovered giraffe as one example. The giraffe, he said, had evolved from an antelope-like creature. As successive generations stretched their necks to eat foliage from trees, their necks became a little longer. They then passed this characteristic on to their offspring. After a long period of time, modern giraffes evolved.

Lamarck's theory incorporated an idea which was common in his day: the inheritance of acquired characteristics. It was generally thought that traits acquired by an individual would be passed along to its offspring. It was believed, for example, that the children of a blacksmith were likely to have unusually strong arms. Today, the idea that such a thing is possible is called "Lamarckism." As we have seen, there was more to Lamarck's theory than that. It depended not only on the inheritance of acquired characteristics, but also on a striving toward higher evolutionary levels.

Today we know that acquired characteristics are not inherited. Changes in the muscle cells of a blacksmith (or in our day, of a weight lifter or football player) cannot affect the ova and sperm that carry the genetic information that will be passed on to succeeding generations. If I spend a lot of time lifting weights (I don't, but I can assume for the sake of argument that I do), this will not make my offspring one whit stronger than they would have been otherwise.

Natural Selection

If the idea of evolution was not original with Darwin, neither was the idea of natural selection. A number of different individuals thought of the idea independently. The theory that evolution was caused by natural selection was clearly stated in the appendix of a book entitled *Naval Timber and Aboriculture*, which was published thirty years before the appearance of Darwin's *On the Origin of Species*. After Darwin's theory became famous, the English author of that book, Patrick Matthew, had the words "Discoverer of the Principle of Natural Selection" printed on his visiting cards.

The English naturalist Alfred Russel Wallace also hit upon the idea of natural selection. He wrote a paper on the subject, and in 1858 sent a copy to Darwin, who was already a famous scientist at the time, asking his opinion. When Wallace discovered that Darwin had been working on a book about the idea for years (he had not yet published anything), he agreed to collaborate with Darwin by writing a joint paper, which was published in the *Journal of the Linnaean Society* in 1858. In 1859, Darwin published *On the Origin of Species*. The first printing consisted of 1250 copies. The entire edition sold out on the day of publication.

If Darwin was neither the originator of the idea of evolution, nor the first person to think of natural selection, why do we speak of "Darwinism" and of "Darwin's theory of evolution"? While others had suggested the idea of evolution by natural selection, it was Darwin who amassed the evidence which showed how plausible the idea was. Darwin had thought of the idea of natural selection in 1838, some twenty-one years before *On the Origin of Species* was published. In the ensuing years, he had studied numerous subjects, such as the breeding of domesticated plants and animals. He realized that, in many respects, natural selection and the artificial selection practiced by plant and animal breeders were similar. The breeders hadn't produced any new species. But they had certainly succeeded in producing genetic alterations. Darwin also studied evidence pertaining to fossils and the geological record, as well as hybridism, morphology, embryology, and the geographical distribution of living species. All these subjects were discussed in his book. Not only did Darwin show that the theory of natural selection was plausible, he also presented so much evidence for the evolution of species that it was no longer possible to doubt its occurrence.

I briefly outlined the idea of natural selection in Chapter 1. Perhaps this would be a good time to discuss it in somewhat more detail. Darwin observed that, in all species, many more offspring are produced than can possibly survive. Even in such a slow-breeding species as the elephant, enough offspring are produced that if they all survived to maturity and produced offspring of their own the world would be overrun with elephants. Second, not all members of a species are the same. This variation ensures that some of them are more likely to survive and produce viable offspring than others. Their hereditary characteristics are therefore the ones most likely to be passed on to the next generation. Thus, over long periods of time, favorable traits will be transmitted and unfavorable ones eliminated. As the environment changes, a species that is adapting itself to its environment will change also. Even if environmental conditions remain relatively constant, a species may change as the individuals that make up the species become more fit. It is this natural selection that causes a species to evolve.

At this point, I should probably say something about the well-known phrase, "the survival of the fittest." This term was coined, not by Darwin, but by the British philosopher Herbert Spencer. Spencer was a Lamarckian, not a Darwinist, and he attempted to apply evolutionary ideas to his theories about human society. Some of these theories, incidentally, were quite harsh. Spencer believed, for example, that no attempts should be made to alleviate malnutrition among the poor. The very fact that they were poor indicated that they were "unfit," and if some of them starved to death, this would work toward the betterment of the human species. In general, he opposed all types of social legislation, including laws regulating hours of work and child labor. Anything that alleviated the struggle for existence, he believed, would work to the detriment of the human species. He even opposed public education, believing that truly superior individuals would rise to the top whatever their environment.

Perhaps I am doing a disservice to Spencer by focusing upon this aspect of his work. Spencer was a kind of forerunner of contemporary scientists who do research in the sciences of complexity. Though Spencer's terminology was somewhat different than today's, he saw evolution as a kind of emergent property which was operative, not only in the biological, but also in the physical world. However, his ideas about "social Darwinism" caused much of his work to be repudiated, and eventually forgotten.

Ironically, the one surviving Spencerian idea that has flourished is that of "the survival of the fittest." This concept is not a very useful one. In the first place, it is somewhat tautological. Who are the "fittest"? They're the ones who survive. Second, it obscures the real essence of the idea of natural selection. If an organism survives but does not reproduce, it will not pass its genes along to the next generation, even though it may otherwise be very "fit." Natural selection, as I have emphasized, depends upon the production of viable offspring.

The Problem of Heredity

At this point, it might be useful to compare Darwin's theory to Lamarck's. As we have seen, according to Lamarck, giraffes had long necks because members of successive generations had been stretching their necks upward in order to reach tree leaves. Spencer thought giraffes have long necks today as a consequence of the stretching that their ancestors engaged in. But Darwin believed this wasn't the case at all. Natural variability that was already present would have caused some animals to have slightly longer necks than others. The giraffes with longer necks would have been more likely to survive and would have passed this characteristic on to their offspring. The change in average neck length from generation to generation would have been very small. Over long periods of time, the necks of the giraffes would have become significantly longer. Darwin's theory was one that emphasized gradualism. Small changes over a large number of generations were all that was required if evolution was to take place.

There was one grave drawback to Darwin's theory. According to ideas that were prevalent at the time, natural selection should not have worked! In Darwin's day, it was generally believed that inheritance was a kind of "blending." For example, if a very tall man married a very short woman, their children would likely be of average height. Similarly, the children of a slave holder and one of his black slaves (remember this was before the American Civil War) would have skin color of an intermediate shade. According to nineteenth century ideas, something similar should have happened in the case of giraffes. If a long-necked giraffe mated with a short-necked one, the offspring would have necks intermediate in size. Over a period of a few generations, the hereditary characteristics that gave rise to long necks would be swamped out. There should have been no slow, progressive trends in either one direction or the other.

Darwin was well aware of the difficulty. Since, in his day, nothing was known of genes or genetics, there was no easy way out of it. As a result, Darwin finally fell back on the Lamarckian idea of inheritance of acquired characteristics. In 1865, six years after the publication of *On the Origin of Species*, he developed a theory he called pangenesis. According to this hypothesis, an organism's cells could produce minute corpuscles called *gemmules* which could affect cells in the reproductive organs. This would produce the variation on which natural selection could act.

Darwin also believed that variations had to arise simultaneously in the majority of the members of a species. Otherwise they would not persist. The blending characteristics of inheritance would have seen to that. Today we know this is not true. A single organism can transmit a useful gene to its offspring. If that gene enhances the probability of survival, then over a period of a number of generations, it can spread throughout an entire population.

Mendel's Peas

In 1856 the Austrian monk Gregor Mendel began a series of experiments in his monastery garden. He obtained some pea seeds of a number of different varieties. All were of the same species, but the plants that grew from them had different traits. Some produced white flowers, while others produced purple flowers. Some of Mendel's pea plants produced yellow seeds, while others had seeds that were green. There were plants with wrinkled seeds, and plants with smooth, round ones. In all, the pea plants varied in some seven different ways.

Mendel crossbred different varieties of plants with one another to see how the various different traits were transmitted from generation to generation. He made exact counts of plants that had this or that trait, and accumulated data that allowed him to find the laws governing inheritance.

I won't describe Mendel's results in detail. It should be sufficient to say that he discovered that inheritance was not a matter of blending. He found, for example, that purple-flowered plants possessed some factor that would cause their offspring to have purple flowers, while white-flowered plants contained a different factor that passed along the white color to their offspring. Although there was often some blending when the two varieties were mixed, the pure traits tended to reappear in subsequent generations. This showed that hereditary factors present in the plants did not disappear, and that there was something seriously wrong with commonly held ideas about inheritance. Mendel didn't call these hereditary factors "genes." That term was invented by the Danish biologist Wilhelm Johannsen in 1909. However, the mechanisms of inheritance are clearly stated in Mendel's work.

In a way, Mendel was rather lucky. Many years later it was discovered that each of the characteristics he studied was controlled by a gene network on a different chromosome of the pea plant. If this had not been the case, he probably would not have been able to obtain such clear cut results. As we have seen, genes operate in networks, and a single gene may be a member of a number of different ones.

This often causes biological characteristics to become intertwined. If the genes for flower color and seed color had been on the same chromosome of the pea plant, there could easily have been a correlation between the two, and this would have muddled Mendel's results. Fortunately, this was not the case.

Mendel's discovery was one of the most important in the history of biology, but it attracted no attention at the time. When Mendel read a paper about his work at a meeting of the local society of natural history, there were no questions and there was no discussion. He sent his results to the distinguished Swiss botanist Karl Wilhelm von Nägel, who showed little interest. von Nägel did offer to grow some of Mendel's seeds, but he never did this, and probably never intended to. Most likely he was simply being polite. In any case, von Nägel didn't bother to answer Mendel's subsequent letters.

In 1865 Mendel published his results in an obscure journal, *Transactions of the Brünn Natural History Society*. This was followed by another paper in 1869. His papers attracted no notice. After he was made abbot of his monastery in 1868, he did little further scientific work. His administrative duties and his increasing girth (it's hard to cultivate plants if you have difficulty bending over) prevented that. When Mendel died in 1884, his work was still unknown.

Then, in 1900, Mendel's work was independently rediscovered by three European botanists, Hugo de Vries, Carl Erich Correns, and Erich Tschermak von Seysenegg. These three scientists had separately worked out the laws of inheritance, and when they searched the scientific literature on the subject, they found that Mendel had anticipated them. To their credit, each of these three scientists gave priority to Mendel and presented their own work only as confirmations of his discovery. The science of genetics had been born.

Genetics versus Natural Selection

When the science of genetics was being developed during the early years of the twentieth century, many scientists believed that Darwin's theory had been superseded. There was no longer any need for this vague notion of natural selection, they thought. Evolution could be explained in terms of genetic inheritance and mutation. Genetics seemed to them to be a superior approach because it dealt with factors that could be quantified.

Furthermore, Darwin had spoken of continuous variation within a species, and of gradual changes over long periods of time. The early evidence compiled by geneticists showed that different genes produced quite noticeable effects. For example, Mendel's pea plants had either white or purple flowers. There was no continuous variation in between. Mutations, the initial evidence indicated, had large effects. It was hard to see how this could be reconciled with Darwin's idea of gradual change.

However, by the 1920s, it had become apparent that many mutations did have small-scale effects, and that combinations of genes could produce continuous variation, such as height in human beings. Matters were not so simple as the early geneticists had believed.

Finally, during the 1930s, the British geneticist J. B. S. Haldane, the American geneticist Sewall Wright, and the British statistician and geneticist Ronald Fisher successfully reconciled genetics with Darwin's theory. They mathematically analyzed mutation rates, reproduction, and other factors, and showed that genetics and the theory of natural selection were not mutually inconsistent. Fisher also investigated gene linkage and developed mathematical methods for dealing with the phenomenon. Suddenly, Darwin's theory had a mathematical foundation.

By 1940, the "neo-Darwinian synthesis" initiated by Haldane, Wright, and Fisher was complete. The idea that natural selection molded evolutionary change was more firmly established than it ever had been before. Even though the units of heredity—genes— were discrete, they could give rise to the continuous variation that Darwin's theory demanded, and evolutionary change could be gradual, as Darwin had insisted.

Is Natural Selection Everything?

All reputable biologists agree that natural selection is the major force in evolution. But is it the only operative factor? Stephen Jay

Gould, for one, doesn't think so. He states quite explicitly that selection "just isn't capable . . . of explaining all major patterning forces in the history of life." He believes that in order to explain long-term evolutionary trends, it is necessary to consider such factors as "the distinctive success of some species versus others," catastrophic extinctions, and the "rapid reorganization of the genome" that may take place when a new species is created. Furthermore, Gould says, there are numerous constraints on the ways in which species can evolve. For example, the same genes regulate the formation of top and bottom surfaces in both insects and vertebrates. Both kinds of organisms build bodies using the same genetic pathways. It is not very likely that the members of some species would suddenly start doing things differently. It is much easier to build upon structures that already exist.

Gould's sometime collaborator, paleontologist Niles Eldredge of the American Museum of Natural History, has expressed similar thoughts. Evolution, he says, is a process that takes place at different scales and levels. In his book *Reinventing Darwin* he says, "Events and processes at any one level cannot possibly explain all phenomena at higher levels." In other words, evolution cannot be explained entirely in terms of its "component parts." When he uses this term, Eldredge is referring to the effects that natural selection has on individual organisms.

An entirely different point of view has been expounded by Richard Dawkins in such books as *The Selfish Gene* and *The Blind Watchmaker*. According to Dawkins, natural selection explains practically all the features of living organisms. Furthermore, he says, the gene is the basic unit in evolution. Individual organisms are nothing but "lumbering robots." They are "survival machines" that are used by genes, which seek to leave as many copies of themselves as possible in future generations.

When Dawkins speaks of the primacy of genes, he does so only for emphasis. His often-used term "gene selection" is really only a metaphor. Naturally Dawkins doesn't think that genes are consciously trying to reproduce themselves. And, as he admits, the ideas that he expounds are really orthodox Darwinism. Natural selection operates at the level of the individual organism. Some survive and reproduce (and thus manage to propagate their genes), while others do not. When Dawkins emphasizes the role of the gene, he is not doing much more than shifting the focus from organisms to the genes themselves.

Gould calls such scientists as Dawkins "hyper-Darwinists" or "ultra-Darwinists." He does not deny that natural selection is the major force in evolution. What he objects to is the idea, expounded by Dawkins and others, that natural selection can account for practically all of the features of any organism. Organisms may have characteristics that are not the result of selection and adaptation to their environments, he says. The "adaptionist program" attempts to explain too much.

Naturally Dawkins disagrees. He downplays Gould's ideas. "Gould seems to be saying things that are more radical than they really are," he replies. "He pretends. He sets up windmills to tilt at which aren't serious targets at all."

So far, I have been discussing the controversy in a rather abstract way. This may make it hard to grasp exactly what the argument is about. It might help to learn about a 1979 paper Gould co-authored with the Harvard evolutionary geneticist Richard Lewontin, "The Spandrels of San Marco and the Panglossian Paradigm: A Critique of the Adaptationist Programme." In it, the two scientists argued that organisms might have characteristics that are not adaptive, but rather accidental. They made use of an analogy which referred to a certain architectural characteristic of St. Mark's cathedral (called *San Marco* in Italian) in Venice.

San Marco has a dome supported by two arches that intersect one another at right angles. This creates four triangular spaces between the arches. There are mosaics of the four biblical evangelists in the spandrels, and images representing the Tigris, Euphrates, Nile, and Indus rivers. According to Gould, the spandrels perform no particular function. They are the "nonadaptive" side effects of the architecture of San Marco. If the dome was to be supported by intersecting arches, the spandrels arose by necessity. They weren't created to house mosaics of the evangelists. On the contrary, they were decorated because there was an empty space to be filled.

According to Gould, the human ability to learn to read and write is a spandrel, as are most of the things that our brains do. Natural selection caused the human brain to get big, he says, "for a small set of reasons having to do with what is good about brains on the African savannahs." But once it got big, it attained a computational power that allowed it to do thousands of things that have nothing to do with "why natural selection made it big in the first place."

Darwin's Dangerous Idea

One might have expected that the controversy would eventually have died down. After all, neither Eldredge nor Gould and Lewontin has denied the great importance of natural selection. Adaptationists like Dawkins are willing to admit there might be some structures not shaped by selection. Furthermore, Gould sometimes used adaptationist arguments himself. He has suggested that the reason locusts appear in thirteen or seventeen-year-cycles is that these are prime numbers. This presumably makes it difficult for predators which might appear every two, three, or five years to track them. If the locusts had, say, an eight-year-cycle, then any predator that appeared every two or four years would always be there waiting for them.

But the controversy didn't fade away. In 1995, Tufts University philosopher Daniel Dennett published a book entitled *Darwin's Dangerous Idea*. In this book, he made the claim that many people feared the implications of Darwin's theory of natural selection. They were presumably afraid that if selection really was a "mindless, purposeless, mechanical process," this could rob life of all meaning and dissolve the illusion of "our own divine spark of creativity and understanding."

Dennett said little about the creationists who would do away with evolution altogether. His fury was reserved for those whom he thought were attempting to water down "Darwin's dangerous idea." His attacks on Gould were especially vehement and lengthy. He charged that Gould was a would-be revolutionary who repeatedly attempted to challenge evolutionary theory, but whose challenges really didn't amount to much. Gould, he said, saw himself as the "Refuter of Orthodox Darwinism." Yet none of Gould's attacks had proven to be "more than a mild corrective to orthodoxy at best." For that matter, Dennett went on, Gould had erred in referring to "spandrels." He really should have called them "pendentives." And spandrels (or pendentives) could have been chosen as an architectural feature because they provided suitable surfaces for the display of Christian iconography. Thus they might be adaptations after all.

It's hard to see how arguments about architectural issues have any bearing on biology. Once Dennett gets going, he doesn't let up. Thus it is not surprising that Gould replied in kind. At first he said nothing. He answered Dennett in two articles published in consecutive issues (June 12 and June 26, 1997) of the *New York Review of Books*. Commenting that "personal attack generally deserves silence by way of response," he explained that he was making an exception in this case because a "demonstrably false charge, if unanswered, may acquire 'a life of its own.'"

In his reply, Gould coined the term "Darwinian fundamentalism" to describe the position taken by Dennett, Dawkins, and other "ultra-Darwinists." He began by noting that Darwin himself had protested against those who would simplify his theory by claiming that natural selection, and only natural selection, caused all evolutionary change. Gould wasn't quoting Darwin to bolster his case; arguments from authority mean little in science. What he was saying was that Dennett and others of his ilk were more "fundamentalist" than Darwin himself.

Gould went on to reiterate his claim that "selection cannot suffice as a full explanation for many aspects of evolution." He added that, in saying this, he was not trying to "smuggle purpose back into biology," as Dennett had implied. These additional causes were "as directionless, nonteleogical and materialistic as natural selection itself." He then replied to Dennett's charges in a manner not unlike that which Dennett had used in his book. Noting that the nineteenth century British biologist T. H. Huxley had been called "Darwin's bulldog" for his vociferous defense of evolutionary theory, he suggested that it was hard to avoid thinking of Dennett as "Darwin's lapdog." After arguing against Dennett's thesis that natural selection can account for everything—or almost everything in evolution—he responded to what he called Dennett's "slurs and sneers" by indulging in a few slurs and sneers of his own. Dennett couldn't get his dates right, Gould said. He had stated that the Cambrian explosion had taken place 600 million years ago. The correct date was 530 to 535 million years ago. For that matter, Dennett couldn't even spell. When he had listed eight of the creatures found in the Burgess shale, he had misspelled three of the names.

Finally, Gould attacked Dennett's attempts to extend Darwinian ideas to areas outside of evolutionary biology (in *Darwin's Dangerous Idea*, Dennett had discussed such topics as human culture, artificial intelligence, and morality). He concluded by again explaining the concept of spandrels, especially as they related to the human brain.

A Real Muddle

At this point, you're probably asking, "Who's right?" Is natural selection all-powerful as Dawkins and Dennett claim? Or is Gould right when he says that there are other important causal factors in evolution? This is not an easy question to answer. Few evolutionary questions are. When theoretical controversies arise in physics they can generally be settled by performing some appropriate experiment. A mathematical theorem can be shown to be incorrect if a fallacy is found in one of the arguments. Matters are not the same in the field of evolutionary biology. Evolutionary change is something that happens over many thousands or millions of years. Though there is a fossil record, we can't go back in time to observe exactly what took place, or determine why it did. The fossil record isn't always that easy to interpret, for that matter. If we find one species was replaced by another at some point in time, it isn't always possible to tell whether this was the result of evolution or whether the older species simply migrated out of the area because of climatic change.

I am a little put off by Dennett's ad hominen style of argument. It is difficult to argue with someone who constantly implies that, if you disagree with him, it is because you are afraid of the ideas that he is expressing. I do admire Gould's willingness to introduce new ideas into evolutionary theory and to attempt to explore its subtleties. This in no way implies that Gould is right and Dennett is wrong. My sympathies lie with Gould. But of course science is built on empirical facts, not sympathies.

As Dawkins and Dennett point out, it is possible to make adaptive arguments about the evolution of certain biological structures. For example, there exist plausible theories of how wings evolved in birds and in insects, and about how the eye evolved in numerous different creatures. There is no way of determining whether or not these theories are actually true, or whether the evolution of wings or eyes depended on selection alone.

Perhaps it would be best to stand back a little. Rather than dwell on the charges and countercharges that have been made, it would be useful to take a deeper look at what Dawkins, Dennett, Eldredge, Gould, and Lewontin are really saying. The members of one camp (Dawkins and Dennett) claim that a reductionist approach is fully adequate. Once one understands the workings of natural selection, there is nothing more to know. Gould and Eldredge (as far as I know, Lewontin hasn't been involved in the controversy) are saying that reductionism is not enough. Evolution is a complex system that cannot be explained in terms of its parts.

I don't mean to present Gould and Lewontin as advocates of the sciences of complexity. They are not (though Gould has spoken favorably of the work of some of the scientists in the field, notably that of Stuart Kauffman). But they do express a similar point of view. When one studies a complex system, it is not enough simply to gain an understanding of the workings of its fundamental parts.

Fitness Landscapes

Like Gould and Eldredge, Stuart Kauffman believes that, although natural selection is powerful, it is not the sole source of order in biology. "Natural selection is always acting," he says. But complex systems have self-organizing properties, and this must also be taken into account. In an essay published in 1995, he summed up his views by saying, "The natural history of life is some form of marriage between self-organization and selection. We must see life anew, and fathom new laws for its unfolding."

As I have previously noted, Kauffman is a theoretical biologist. He doesn't study the fossil record, or organisms in the natural world. Instead, he relies on computer simulations and theoretical arguments. To understand his ideas, it is necessary to backtrack a little and to review a previously discussed concept, one introduced by Sewall Wright in the 1930s.

Wright invented the idea of fitness landscapes. They provided a new way of looking at evolution, and the idea has been used extensively, not only by Kauffman, but also by such "ultra-Darwinists" as Richard Dawkins. A fitness landscape is just what the name implies. It is an imaginary landscape filled with hills, peaks, and valleys. The peaks are regions of high evolutionary fitness. As succeeding generations of organisms become better adapted to their environment, they climb up to higher altitudes; natural selection draws them there. An organism that lived in a valley would be so poorly adapted to its environment that it would probably not survive. "Low-altitude" populations are quickly weeded out.

Fitness landscapes provide a visual way of thinking about genetic change. Natural selection is the force that drives populations toward peaks. Natural selection acts only on individuals, not on entire species. If we have a population of better adapted individuals, we can say that the species is well adapted too. That is what fitness is.

Kauffman has done some interesting studies of fitness landscapes. He finds, for example, that if the landscape is constructed randomly, for example by generating random numbers in a computer (remember that Kauffman works with computer simulations), little evolution will take place. A random landscape is one that will have a lot of peaks, sheer cliffs, and precipices. It has this appearance because there is no correlation between the altitude of any given point and one that is nearby. In such a landscape, a population that has evolved to a local peak will most likely have no way of getting to much higher peaks farther away. There will be too much rough terrain in between, and its climb uphill will be halted before its adaptation to its environment has become optimal. Mutation could presumably allow it to jump across chasms, but there are so many peaks in such a landscape that its chances of finding the highest peak, or one near it in height, is nil. In effect, it will remain trapped at some modest height.

Kauffman says his models show that a fitness landscape that is too smooth will not work very well either. The population will probably find a reasonably high peak. Mutation may cause it to diffuse away down the adjoining slope and drift away toward the lowlands. The gentle slope that surrounds the peak cannot prevent this. Thus the population can lose the adaptations it has so painfully acquired.

Real landscapes are neither completely random nor perfectly smooth. For example if one is at a high altitude in the Alps, there will probably be a number of other high peaks in the vicinity. If you walk or climb in a randomly chosen direction, you are not likely to suddenly find yourself in the lowlands again. Real mountain peaks are often found in groups. The most famous Alpine group is the trio of mountains known as Mönch, Eiger, and Jungfrau (German for "Monk," "Ogre," and "Virgin"), which are all in the same vicinity. If you move away from one of these mountains, you can easily find yourself walking up a slope that leads to another.

Kauffman calls such landscapes "correlated." He sometimes refers to them as "rugged" to distinguish them from landscapes that are smooth or truly random. And now we come to the interesting part. Kauffman has performed computer "experiments" which seem to indicate that the character of a fitness landscape depends upon the interconnectedness of a population's genes. Kauffman's research shows that if every gene could act upon or be acted upon by every other, the interconnections would be at a maximum, and a random fitness landscape would result. If there are no interconnections, if each gene acts independently, the landscape will be smooth. Computer studies, Kauffman says, show that a good, correlated landscape will be produced when each gene interacts with a small number of others.

If Kauffman's analysis is correct, then it has striking implications. If natural selection needs certain kinds of fitness landscapes in order to work, then it may have acted upon gene networks by favoring those with an optimal degree of interconnectedness. In other words, evolvability may be a product of evolution too. After all, the gene networks did not acquire the characteristics they exhibit for no reason. They are also products of selection.

At first, this conclusion seems to contradict Kauffman's assertion that there are other factors in evolution besides natural selection. But we have not reached the end of his argument. We have to consider how selection got a start in the first place. As Kauffman says in his book, *At Home in the Universe,* "Were cells and organisms not inherently the kinds of entities such that selection could work, how could selection gain a foothold? After all, how could evolution itself bring evolvability into existence, pulling itself up its own bootstraps?"

He then answers his own questions by appealing to an idea that is encountered again and again in the sciences of complexity: self-organization. Self-organization may have initially generated the kinds of structures that could benefit from natural selection. In other words the fact that complex systems have emergent properties may have led, not only to the creation of the first living organisms, but also to their evolvability. According to this view, natural selection causes species to evolve, but self-organization creates the plasticity that makes evolution possible in the first place.

What is Dennett's response to all this? In *Darwin's Dangerous Idea,* he says of Kauffman, "Many have heralded him as a Darwin-slayer." But then he concludes that Kauffman is nothing of the sort. His view seems to be that Kauffman is a kind of "meta-engineer" who has added something to Darwinian theory by discovering some of the basic rules of design for living organisms and their genetic networks. Dennett is also willing to discuss Kauffman's work in a mildly unfavorable light. In a commentary on an interview with Kauffman, conducted by New York literary agent John Brockman and published in 1995, Dennett said that Kauffman "feeds the yearning of those who don't appreciate Darwin's dangerous idea. It gives them a false hope that they're seeing not the forced hand of the tinkerer but the divine hand of God in the workings of nature." In *Darwin's Dangerous Idea,* which was published the same year, he paid Kauffman a left-handed compliment by saying that Kauffman had originally styled himself a radical heretic, but then had moderated his rhetoric, in contrast to Gould, who had gone "from revolution to revolution."

A Throne Covered with Crocodile Skin

In 1995, two University of Sheffield scientists, Ricardo Colasabti and Tash Loder, described the opening of a meeting of the Linnean Society (named after the eighteenth century Swedish botanist Carl von Linné—his Latin name was Linnaeus) in London in a posting on the internet with the following words:

> The evening started with all the pomp that only England can (or want to) muster. The Chair of the Linnean Society, with no trace of amusement, announced new members whilst sat on a throne (which appeared to be covered with crocodile skin) wearing a large three-cornered hat which would not have looked out of place in a low-budget television drama.

The primary purpose of the meeting, however, was not to induct new members, but to provide a setting for a debate between Stuart Kauffman and the famed English geneticist John Maynard Smith (one of the scientists whom Gould includes among the "Darwinian fundamentalists"). Kauffman presented his theories about complexity and fitness landscapes in an hour-long talk. Maynard Smith then spoke for an hour. Finally there was a joint half-hour question period.

Although Maynard Smith agreed with some of the ideas expressed in Kauffman's talk, such as the need for a theory of complex phenomena, he expressed skepticism about Kauffman's theories. They were not sufficiently grounded in reality, he said. Other models could also produce similar complex results "when flexed in the correct manner." During the question period, he said, "My problem with Santa Fe (the Santa Fe Institute, where Kauffman does his research) is that I can spend a whole week there . . . and not hear a single fact." Kauffman responded, "Now that's a fact!" The remark was greeted with loud laughter.

The remarks that were made at the Linnean Society meeting outline some of the difficulties faced by those who wish to deal with issues involving complexity. Nowadays scientists generally agree that complex systems do exhibit emergent behavior which cannot be explained by analyzing them into their component parts. However, complexity is a relatively new area of research, and it is not always easy to relate theoretical studies to experimental work. In many cases, it is not even very clear what experimental studies could be performed. Thus it is not easy to tell how much validity theoretical models (such as those that Kauffman has produced) have.

It might be worthwhile to compare work done in the sciences of complexity to the physics of the early decades of the twentieth century. This was a period during which a number of new theories, such as quantum mechanics and Einstein's special and general theories of relativity, were formulated. Unlike the sciences of complexity, physics had existed as an experimental science for a long time. Thus Einstein could point out that his new theory of relativity had been designed to explain known experimental facts, and he could suggest additional experiments that might be performed. The situation with respect to theories such as those of Kauffman is entirely different. Although they seem logical and appealing, it is not yet so easy to see how they might be tested.

The scientists who work in the sciences of complexity understand this. As I write this, scientists at the Santa Fe Institute are seeking support from NASA for a program that would combine theoretical and laboratory work on replicating biological molecules. Kauffman and others, inspired by Reza Ghadiri's demonstration that peptides can replicate themselves, wonder whether it might not be possible to create autocatalytic sets in the laboratory. "We're within a decade or so of making communities of self-reproducing systems," he has said, "and we're going to invent a whole new body of theory of emergent biological phenomena."

It's somewhat more difficult to see how Kauffman's theory of gene interaction and fitness landscapes could be tested. The idea of a fitness landscape is, after all, something of a metaphor. No one knows how to measure fitness numerically, or how to assign numbers to the height of peaks in a landscape. This can easily be done in a computer model, but in the real world matters become somewhat more complicated. Furthermore, adaptation is something that takes place over many thousands or millions of years. The only way to observe evolutionary changes is by examining the fossil record. The fossil record is incomplete and often difficult to interpret. Recall that we have already seen one example of this. Climatic change may cause species to migrate to different habitats. Thus if a "new species" suddenly appears in some sedimentary rock in which fossils are preserved, it is often not always possible to tell whether it suddenly evolved, or whether it simply moved from a region in which few or no fossils were formed.

I suspect experimental evidence that either supports or casts doubt on Kauffman's theory of autocatalytic sets may be obtained as soon as the first decade of the twenty-first century. However, I won't be surprised if his theory of genetic networks and the structure of fitness landscapes will be argued about for quite some time. To be sure, scientists will learn more about the interactions between individual genes. This, in itself, is not likely to be enough to confirm or disconfirm the theory. At this point, it is just too far removed from experiment. If that situation changes, the change is likely to be gradual.

Something similar can be said about many other theories in the sciences of complexity. Maynard Smith may have had a point when he said that he could go to the Santa Fe Institute and not learn a single fact.

5

ARTIFICIAL LIFE

E volutionary biologists are often skeptical of the value of computer models. After all, life is a very complicated phenomenon, and when evolutionary processes are modeled on a computer, a number of simplifying assumptions have to be made. There is really nothing wrong with doing this. As I have remarked previously, scientists almost always make a number of simplifying assumptions when they begin to study a problem. But the biologists often remain unconvinced, feeling that simulations cannot mirror the full complexity of life. A good example of this attitude can be found in a remark once made by the noted Ukrainian–American geneticist Theodosius Dobzhansky, who compared computer modeling to masturbation. It was often enjoyable, he said, but it was no substitute for the real thing.

Nevertheless, computer simulations often exhibit phenomena that bear an eerie resemblance to those that are observed in reality. As I mentioned in Chapter 1, biologist Tom Ray succeeded in creating electronic "organisms" that could reproduce, mutate, and evolve. Of course they did all this electronically, existing only inside a computer. When Ray ran his program, "Tierra," he discovered that the changes in the creatures in his electronic world mimicked many different kinds of evolutionary phenomena. Not only did his creatures compete with one another for "food" resources and electronic survival, at one point parasites evolved in his system, and preyed upon the electronic organisms that were the original inhabitants of Tierra. The latter responded by developing defenses against the para-

sites. Ray also found that his artificial organisms would often change
little over the course of numerous generations, and then experience
sudden bursts of evolutionary activity. The latter phenomenon bore an
uncanny resemblance to what biologists call punctuated equilibrium.

I'll have more to say about Ray's artificial life experiments later
on. But first I want to provide a little background by discussing the
theory of punctuated equilibrium, which was proposed by Stephen
Jay Gould and his colleague Niles Eldredge in 1972. Gould and Eldredge believed it was a mistake to assume that natural selection was
a process that caused only gradual changes in organisms over long
periods of time. On the contrary, the fossil record indicated that
species typically experienced little change over periods of many millions of years. These periods of stasis were punctuated with sudden
bursts of evolutionary activity during which new species suddenly
appeared.

When Darwin published *On the Origin of Species* in 1859, he
claimed that natural selection would cause species to change slowly
over long periods of geological time. This idea was not supported by
the empirical evidence that was available in his day. Nearly every
paleontologist who reviewed Darwin's book pointed out that this
was not what the fossil record showed. Gradual evolution, with a lot
of intermediate forms, was not what they saw. Darwin's view was in
conflict with observations.

Darwin attempted to meet the difficulty by pointing to gaps in
the fossil record. Many intermediate forms might appear to be missing, he admitted. Nevertheless, paleontologists had found species
that appear to be descended from earlier ones. Darwin asserted that,
as more evidence was accumulated and the fossil record became
more complete, the missing details would be filled in. Paleontologists would discover the slow gradations that his theory seemed to
require.

As it turned out, paleontologists never did discover as many
intermediate forms as Darwin thought they would. They found species that changed gradually over time, but these changes often did
not appear to be very significant. After millions of years of "evolution," many species were very much like their ancestors. In some

cases, the changes were so small that it was said by some scientists that they could not be attributed to natural selection!

You should not imagine that this casts doubt on the fact of evolution itself. For example, we are clearly descended from *Homo erectus*, and the modern horse is a descendant of small three- and four-toed horses that roamed the plains of North America millions of years ago. Similarly, there is no doubt that whales are descended from land mammals that returned to the sea. Paleontologists have found fossils of primitive whales that could still move about on land.

However, there is a problem, that of determining why evolution has not proceeded as Darwin imagined it should. *Homo erectus* survived as a species for 1.3 million years. During that time, there was little or no enlargement of its 1000 cubic centimeter brain (the average brain size of modern *Homo sapiens* is 1300 to 1400 cubic centimeters). Modern whales and bats are not so very different from what they were when they evolved 55 million years ago. And there exist insects that have not evolved significantly in 100 million or more years. Natural selection is supposed to induce slow but inevitable change. Why then are there so many cases where it seems not to have had that effect?

To be sure, there is nothing in Darwin's original theory, or in the modern neo-Darwinian synthesis (i.e., the synthesis of Darwinian theory and genetics) that implies that evolution must proceed at a constant pace. Biologists have traditionally thought that when environmental conditions remained constant, a species would change very little. If a population is already adapted to its environment—if it already occupies an adaptive peak in the fitness landscape—change is only likely to lead to lower fitness. However, it was thought that when the environment changed, the fitness landscape would change too. Populations would be forced to climb to new peaks or to reclimb ones that had shifted their position.

There are complications. Matters are not always as simple as this. If environmental conditions change, members of a species may simply migrate in response. If the climate becomes warmer, for example, they may move farther north. Incidentally, this applies to plants as well as animals. They can "migrate" by spreading seeds.

Thus an entire ecosystem—including birds, mammals, plants, insects, and fungi—can move to an area where conditions are more favorable. It may be that such an environment cannot be found. In such cases, a species is likely to respond by becoming extinct.

Even though it was admitted that such exceptions might exist, the "ultra-Darwinist" view was not altered. Evolution happened because populations adapted themselves to their environments. And they did this in a slow, gradual way, as the theory predicted. It was difficult to imagine how anything else could be the case.

Paleontologists versus Geneticists

Nevertheless, there seemed to be much more stasis in the fossil record than conventional theory could account for. How did evolutionary biologists respond to this fact? For the most part, they ignored it. The neo-Darwinian synthesis was fabricated, not by paleontologists who saw stasis in the fossil record, but by geneticists who created mathematical models of gene frequencies and evolutionary change. Theory tended to regard evolution in purely genetic terms.

When I say all this, I don't mean to denigrate the work of the geneticists or of the biologists who elaborated upon their work. Discovering the mathematical details of the manner in which natural selection could operate was a great achievement. Furthermore, it was perfectly possible to view much of the gappiness in the fossil record as an illusion. After all, when a new species suddenly appears in a fossil bed, it is not always possible to conclude that it evolved directly from older specimens. It may simply have moved into the area. In such a case there can be no intermediate forms. If they fossilized at all, they did so somewhere else. Fossilization is a very chancy thing. I don't know what the probability is that any individual organism will become a fossil. Nobody does. But we do know that the chances are very, very small. It is not very likely that either your bones, mine, or those of anyone we know will wind up being exhibited in a museum 10,000 years from now. The chances are no better that this will happen to your sister's goldfish, my cat, or many of the wild creatures that still exist in our world. The fossil record that paleontologists study is, by its very nature, incomplete. Every fossil that is found is the result of some lucky (lucky for us, not for the creature that died) event.

Many of the gaps in the fossil record can be explained by looking at the manner in which new species are created. First, consider what will happen if there is some genetic change in a few members of a population. The change may be so unfavorable that these individuals will not survive. Or if they do survive, they will interbreed with other members of the population. Eventually some genetic change may spread throughout the population. There will still be only one species, just as there was before, and little or no evolution will have taken place.

However, members of a species can become geographically isolated. For one reason or another, a population may become separated from the remainder of the species by a geographical barrier, such as a desert or a mountain range. Natural selection will then cause the isolated population to adapt itself to its environment. Over time, it may very well evolve into a new species. This is especially likely to happen if it is a relatively small population. Favorable mutations spread through small populations quickly.

It might be best to illustrate this concept with a specific example. Suppose a certain animal lives in a climate that is becoming drier. Let us further suppose that the climate becomes so dry that two populations of the species become separated by a desert. The population on one side of the desert may evolve more rapidly either because it is smaller, or because it must adapt to new environmental conditions. Then, over a period of tens of thousands of years, the climate becomes wetter again, and the two populations once again come into contact. If the daughter species has become better adapted than its parent, it may very well supplant the latter as it competes for survival in the same ecological niches. If fossilization only occurs in the area where the parent species has been living, paleontologists will never find any intermediate forms. They will find only the parent and daughter species in the geological strata that they study. The relationship between the two will probably be clear; it may be easy to tell that one species descended from the other. But paleontologists will never find examples of what was in between.

Don't Bother the Grownups

The net result of all this is that paleontologists were often confronted by a fossil record that was often difficult to interpret. Mathematical geneticists seemed to have a theory which gave a complete
explanation of evolution by natural selection. So perhaps it is not
very surprising that the views of the latter group became dominant.
In many cases they did not even want to consider any speculations
that paleontologists might have over the mechanisms that had
caused evolution to take place. As the geneticist John Maynard
Smith put it, the "attitude of population geneticists to any paleontologist rash enough to offer a contribution to evolutionary theory
has been to tell him to go away and find another fossil, and not to
bother the grownups."

In 1972, Niles Eldredge and Stephen Jay Gould did offer a contribution to evolutionary theory, one that had to be taken seriously. As
we shall see, the geneticists and "ultra-Darwinists" would eventually decide that Eldredge's and Gould's contribution was only a
minor one. Before I discuss the reaction to the theory of punctuated
equilibrium, it might be best to tell the story of how the theory came
about.

During the 1960s, Eldredge began a study of some long-extinct
marine animals called trilobites. He hoped to document the gradual
evolutionary changes that conventional evolutionary theory said
should have taken place. Instead, he found that his fossils exhibited a
distressing sameness over vast periods of geological time. He would
pick out a trilobite species, and find that it changed little over periods
of 8 million years. To be sure, there were some differences between
the earlier and later specimens, but they didn't seem to be very
significant. As far as he could tell, evolution seemed to have been
going nowhere.

Eldredge's initial reaction was one of frustration; his research
project wasn't yielding any results. Eventually he began to realize
that he had rediscovered a phenomenon that had been known to the
paleontologists who had been contemporary with Darwin. Species
were very stable entities. Stasis, not gradual change, was the norm.
Paleontologists had adopted the theory of gradual evolutionary

change, and had held onto it, even when they found evidence to the contrary. They tried to interpret fossil evidence in terms of accepted evolutionary ideas.

It is not surprising that they should have done so. Scientists generally interpret empirical facts in terms of existing theory. If they did not do this, it would be impossible to do science at all. If a new theory were created to explain every new experimental result or empirical observation, the result would be chaos. Thus scientists generally try to fit new results into existing theoretical frameworks. It is only when a large number of anomalous results are obtained that existing theories are questioned and then modified or overthrown.

You should not imagine that Eldredge's findings cast any doubt on the theory that evolution proceeded by natural selection. His observations showed nothing of the sort. He had only observed that certain species remained static for long periods of time. Species had obviously evolved, just as Darwin said they had. But it was becoming apparent that evolution might not be the slow and steady process that Darwin had envisioned.

Gould's Snails

Stephen Jay Gould is known primarily as a writer on evolutionary topics. As a scientist, he specializes in the study of fossils of snails. At about the same time Eldredge was observing stasis in lineages of trilobites, Gould was seeing the same thing in the snails he studied. New species seemed to appear abruptly, and to remain the same for long periods of geological time. It so happened that in 1971 Gould was asked to contribute a paper to a book that was to be titled *Models in Paleobiology*. He was asked to write on the subject of speciation (the evolution of new species), but this was not the topic he would have preferred. So he asked Eldredge to write the paper with him. The two paleontologists wrote "Punctuated Equilibria: An Alternative to Phyletic Gradualism," and it was published in 1972.

In this article, Eldredge and Gould claimed that new species did not arise "from the slow and steady transformation of entire populations." On the contrary, they said, most evolution took place when new species arose "very rapidly in small, peripherally isolated local populations." They went on to state that, "the great expectation of insensibly graded fossil sequences were a chimera." Sudden breaks in the fossil record represented something real.

In other words, most evolution took place when new species were created. For a short time, evolution proceeded very rapidly. It settled down again once the surviving species had adapted themselves to their new-found ecological niches. According to Eldredge and Gould, evolution is not something that took place over periods of many millions of years. Adaptation over periods of tens of thousands of years seemed more likely. When the fossil record was examined, this led to the appearance of sudden changes. After all, in geological terms, a period of a ten or even a hundred thousand years is an "instant," compared to the many millions of years represented by the rock strata that one is examining.

Eldredge's and Gould's theory seemed to imply that, most of the time, natural selection acted to keep species relatively stable. It only launched into action when a new species arose and had to adapt itself to local conditions. Furthermore, not all new species would survive. Those that were less well adapted were likely to become extinct. Eldredge and Gould called this "species selection" or "species sorting." Sometimes entire species competed for survival; natural selection was not simply a matter of the reproductive success of individual organisms.

Revolutionary Theory or Minor Correction?

When the theory of punctuated equilibrium was reported in the news media, some creationists were quick to make the claim that doubt had been cast on the fact of evolution itself. Nothing could be further from the case. Eldredge and Gould never denied that evolution had taken place, or that natural selection was its main cause. They had simply pointed out that evolution tended to move in spurts with long periods of quiescence in between. They were only elaborating upon an accepted theory, pointing out that evolution did not proceed in quite the manner that most evolutionary biologists assumed.

The "ultra-Darwinists" have often attempted to downplay Eldredge's and Gould's ideas. For example both Richard Dawkins and Daniel Dennett have claimed that the idea of punctuated equilibrium is, at best, only a minor correction to evolutionary theory. If evolution takes place over tens of thousands of years, rather than millions, they say, it is still gradual. The idea that natural selection causes slow changes hasn't been overthrown at all.

When authors such as Dawkins and Dennett say this, they may be missing an important point. In pointing out that evolution may be associated with speciation, Eldredge and Gould showed that it may operate at a higher level of complexity than had previously been thought. In their view, evolution is not simply a matter of individuals competing with one another for survival. When one species splits into two or more, not all of the species may survive. They may very well find themselves competing with one another. Furthermore, some evolutionary lineages may speciate more easily than others, evolve more rapidly, or become extinct more readily. One doesn't have to look far to find examples of this. Both rats and elephants have been around for about 50 million years. The rat has a shorter life span and reproduction time. Thus there have been more generations of rats than elephants during this period. One would expect, then, that the rat has evolved more. In fact, the opposite is the case. It is the elephant which exhibits the greater amount of evolutionary change.

In other words, the reductionist approach of the geneticists, who sought to explain everything in terms of gene frequencies and competition between individual organisms, was really not sufficient. Evolution was a complex process that often operated at higher levels than they liked to imagine.

Having said that, I want to reiterate that I am not using the term "reductionist" in a negative sense, as it so often is these days. Modern science has been successful precisely because it has taken a reductionist approach. Physics is based upon an understanding of the basic constituents of matter and the interactions between them. Alchemy was "holistic." Chemistry, by comparison, depends upon a reductionist understanding of the chemical elements and of the bonds that form between atoms. It was a reductionist approach that led to the discovery of the nature of DNA. Reductionism has enabled scientists to work out the details of the functioning of cells. It is safe to say that, without reductionist techniques, science would be pretty much as it was in 1850. A great deal is said about the virtues of "holism" these days. But I, for one, wouldn't want to give up the great quantities of knowledge that reductionist science has given us.

Reductionism isn't everything. Scientists are beginning to understand that phenomena often cannot be fully understood unless they are viewed at different levels of complexity. The whole is often more than the sum of its parts. As I pointed out in Chapter 1, understanding the properties of a water molecule is not sufficient to allow us to explain why a little whirlpool forms when the plug is pulled in a bathtub. Similarly, Eldredge and Gould seem to have shown that an understanding of the genetic makeup of individual organisms and the mechanics of natural selection is not sufficient to explain everything that happens in evolution. Evolution seems to operate at a number of different levels.

Eldredge's and Gould's outlook is especially interesting because it did not arise out of work in the sciences of complexity. It was expressed by two scientists who had been studying evolutionary phenomena in a traditional manner. And yet the conclusions that they reached are remarkably similar to those of the scientists who study complex systems. Eldredge and Gould are not theoretical scientists who, like Kauffman, try to understand nature by creating computer simulations. They are paleontologists who gather and study fossils, and who sometimes use fossil evidence to shed more light on the workings of evolution. When they formulated the theory of punctuated equilibrium, they did not do so because they had preconceived ideas about the behavior of complex systems. On the contrary, they came to the conclusion that evolution operated at more than one level of complexity because it seemed to be the only way to explain the observed facts.

Life inside a Computer

In 1989 Thomas Ray was talking to some scientists at the Santa Fe Institute about his idea of creating artificial, evolving life inside a computer. It would never work, they told him, or at least it would be very difficult. Computer programs were too "brittle." If a mutation caused an instruction to change, the most likely result would be that the program would cease to function. And, they added, he would have to set up his experiment inside a virtual computer. After all, he was contemplating creating computer virus-like organisms. If he succeeded, and if any of them escaped, they would wreck havoc on other systems. Unlike ordinary computer viruses, they would have the ability to evolve, and thus would be very difficult to eradicate.

"What's a virtual computer?" Ray asked.

A virtual computer was an emulation of a computer inside a real computer, he was told. Creating a virtual computer was something like making a computer model of an automobile. The model had the same properties as a real car, but it existed only in the computer software. If Ray's artificial organisms lived inside a virtual computer, they could never escape because they would look like nothing but streams of data to the real computer.

To many people, all this might have seemed discouraging. But Ray was a biologist, not a computer scientist. The people to whom he had been speaking had much more experience with programming than he did. He went ahead with his project of creating artificial life anyway. He created an artificial world which he named "Tierra" (Spanish for "earth"; he later said that he wanted to avoid the associations evoked by the term "Gaia") and injected it with an electronic organism which he called the Ancestor. Compared to real biological organisms, the Ancestor was not very complicated. Where the simplest bacteria have many hundreds of genes, the Ancestor had just three. And it was only eighty computer instructions long. It was designed to be able to replicate itself, to mutate, and feed on "energy" provided by the computer's central processing unit (CPU). In order to prevent overpopulation, Ray included a subprogram that he called the "reaper." The reaper performed the function of keeping the population down to a reasonable size by killing organisms when they reached a certain age. Some were given the opportunity of lingering on little longer than others, however. If evolution did occur, the reaper would allow the fitter organisms to live on for a while. However fit they became, none would be immortal.

Ray really didn't expect much in the way of results. The advice he had received had convinced him that it would probably take years to complete his project, if he could accomplish his intentions at all. The response that he got from other scientists was hardly encouraging. When he presented a seminar on his ideas to his colleagues at the University of Delaware, he was "virtually laughed out of the room," as he later described the experience. At least he could make some attempts. Setting up the experiment would give him a better understanding of the problems that he would face.

To Ray's surprise, his artificial world sprang into life immediately. His artificial organisms began to proliferate quickly. Initially, Tierra contained nothing but copies of the Ancestor. Almost at once, Ray's electronic organisms began to evolve. The first newly evolved organism was one that contained seventy-nine instructions, as compared to the Ancestor's eighty. Its slightly shorter length allowed it to replicate more rapidly, and its population soon exceeded that of the Ancestor. Then even smaller organisms appeared. Eventually, an artificial creature with only forty-five instructions evolved. With such a small number of "genes" it was not able to replicate on its own. That process required a certain number of instructions, which Ray estimated to be somewhere in the low sixties. The new, small creature was a parasite which made use of the reproductive machinery of its host in a manner analogous to biological viruses, which make use of the biological machinery of the cells they infect.

The parasites proliferated rapidly, but the hosts began to evolve defenses against parasitism. These proved to be so effective that the smaller creatures were eventually eliminated. This was not the end of the story. As evolution progressed further, Ray's creatures learned to use the "energy" they obtained from the computer's CPU more efficiently and to enter in symbiotic relationships with one another. The symbiotic organisms could not reproduce on their own. Instead they shared pieces of computer code to carry out replication. This led, in turn, to the evolution of a new kind of parasite which inserted itself between two cooperating organisms and stole the genetic information necessary to reproduction as it passed from one creature to the other. Thus the new parasite was able to reproduce successfully even though it was significantly shorter than its parasitical predecessors. It contained only twenty-seven instructions, as compared to the eighty possessed by the Ancestor.

How did Ray manage to succeed when everyone thought he would fail? This may be because he was a biologist and not a computer scientist. Since he didn't know how such a program was "supposed" to be written, he modeled the Ancestor on biological organisms rather than on previously developed software. Furthermore, he took the warnings about the "brittleness" of traditional computer programs to heart. This was not difficult to do, since he was imitating biological models.

Two Kinds of Evolution

In both the "real" and computer worlds, biological parasites are simpler organisms than their hosts, and the hosts often do evolve defenses against them. This latter fact hasn't eliminated parasitism; as I noted previously, there are more species of parasites on our planet than there are host species. Biological organisms do share genetic codes. Bacteria often engage in a process called conjugation during which one bacterium will inject bits of DNA into another. This is one of the things that causes resistance to antibiotics to spread so rapidly. If one bacterium acquires a genetic resistance to a drug, it can pass on this resistance to other bacteria, as well as to its offspring. However, there was a significant difference between Ray's parasites and those in the biological world. There exist no parasites which attempt to intercept genetic material as it is passed from one organism to another. Unlike some of Ray's artificial creatures, bacteria do not exchange genetic material in order to reproduce. They do this by dividing in two.

There was one respect in which Ray's artificial world bore an uncanny resemblance to the real one. Evolution tended to take place suddenly, after long periods during which nothing much seemed to be happening. Tierra could be very stable for periods corresponding to millions of computer instructions. And then, suddenly, new artificial organisms would evolve, and the genetic diversity would increase. Something closely resembling punctuated equilibrium had been observed happening inside a computer.

I won't speculate much on the reasons why two entirely different forms of life—one real and one artificial—should exhibit similar evolutionary patterns. This fact does suggest that there may be some deep organizing principle at work. As we have seen repeatedly, complex systems exhibit behavior that cannot be predicted from knowledge of the systems' components. There is no reason why, in some cases, the same organizing principles could not be at work in both biological and artificial worlds. Of course I am on tenuous ground when I suggest this. Whether or not there exist any such "laws of complexity" is one of the questions that scientists have not yet answered.

The Evolution of Complexity (Again)

Ray's artificial organisms exhibited no tendency to become very complex. If anything, there was a trend toward the evolution of smaller, simpler organisms, which were able to use the energy in the CPU more efficiently. Their shorter length gave them an advantage because they reproduced more rapidly.

Ray's experiments had shed no light on the evolution of complexity either in an artificial or in the real biological world. Naturally, Ray realized that this was the case, and he began to wonder if there might not be ways to alter his artificial world in such a way as to induce greater complexity to appear. As I write this, Ray is working on a project in which artificial life will live and evolve in a much richer and more complex environment. Ray hopes that this will cause some very complex artificial life forms to appear. This is a subject to which I will return a little later. At the moment, however, there is more to be said about possible parallels between Ray's work and the ideas of Eldredge and Gould.

The fact that there was no trend toward increasing complexity in Tierra does not imply that evolution in an artificial environment is so very different from the evolution of life on earth. There may be more similarities than appear at first glance. As we saw in Chapter 3, Gould has argued that there is no trend toward greater complexity in evolution. If there are many millions of living species, chance alone will ensure that some are very complex, while the great majority remain simple. If he is right, the kind of evolution observed in Tierra is pretty much what one would expect.

There exists empirical evidence which seems to indicate that Gould's idea could be correct, and that the pattern of evolution observed in Tierra mimics that which is seen in the biological world. I remarked previously that it is probably impossible to create a measure of the complexity of living organisms. However, it is possible to select some specific anatomical feature of an evolutionary lineage of organisms and to determine whether this feature has become more complex during the passage of time. This is precisely what was done by University of Michigan biologist Daniel McShea during the early 1990s. McShea noted that some animals have spinal columns which are more complex than others. The vertebrae of a fish, for example, tend to be pretty much alike. Yet in a typical mammal, there are fewer vertebrae, but they are more differentiated. For example, the neck bones do not look like the back bones, or the bones which support the pelvis.

You should not conclude that this implies that a fish is less complex than a mammal. Fish and mammals, which generally live in very different environments, have simply evolved in different ways. A study of the differentiation of vertebrae really doesn't help us create a general definition of "complexity." However, as McShea realized, making measurements of the characteristics of mammal vertebrae over the course of evolutionary time can at least provide information about what has happened to some arbitrarily chosen feature. Evolution is a complex phenomenon, and broad generalizations must often go untested. Nonetheless, it is possible to study a kind of "micro evolution" and reach some very definite conclusions.

McShea measured six different characteristics of the vertebrae of five groups of fossil mammals that lived over periods of many millions of years, and he compared them to the vertebrae of their living descendants. He found that there was no trend toward increasing complexity. In many cases the descendants did not differ significantly from their ancestors. In cases where there were significant changes, there were actually more decreases in complexity than increases (but the difference was not statistically significant).

McShea's study was a little more complicated than I have made it seem. Biological features cannot become too simple or they will cease to perform any function. Ray's artificial organisms had to have a certain minimal size if they were to be able to reproduce. This was as true of parasites as it was of nonparasitical organisms. Similarly, a spinal column cannot become too simple; if it did, it would cease to perform its function. In other words, there are barriers which prevent complexity from decreasing too much, while there is always plenty of room for increasing complexity on the other side. This fact caused McShea to distinguish between "passive" and "driven" trends. I won't bore you with all the details, however. His finding it necessary to make some fine distinctions does not negate his conclusion that natural selection (at least when one concentrated on this particular anatomical feature) did not cause a trend toward either increasing or decreasing complexity.

McShea thinks that there may be certain evolutionary trends favoring *decreasing* complexity. He has recently been testing the theory that the cells of multicellular organisms are simpler than those of single-celled creatures. The idea is that a cell within a multicelled body does not have to do all the things that a free-living cell has to do. "People are sophisticated," McShea says, "but compared to a paramecium, their cells are simple."

Which brings us back to Tom Ray and his artificial life experiments. If his organisms tended to become simpler, not more complex, this does *not* necessarily mean that they exhibited evolutionary behavior unlike that seen in the real biological world. McShea's theory about multicellular organisms may turn out to be relevant to Ray's work too. He is currently working with "multicellular" artificial organisms in the hope that they will prove to be more likely to show a trend toward increasing complexity than his original "single-celled" creatures did. If he succeeds, it will be interesting to see whether or not the "cells" of his evolving computer organisms become simpler, as McShea thinks they should.

Is Artificial Life Really "Life"?

Even Gould admits that, if enough different species evolve, some of them are bound to be very complex, even if the majority remain simple. If this were not the case, there would be no Stephen Jay Gould to write about them. This raises the question of how complex artificial life is likely to get, and leads to some other very interesting questions as well.

Perhaps it would be best to consider the simpler questions first. For example, should a computer organism that feeds on CPU "energy" in its environment, and which reproduces, mutates, evolves, and dies inside a computer be considered to be "alive"? I suppose the answer to that depends upon one's definition of the term "life." However, I think that we probably have to admit that Tom Ray has come closer to creating life than scientists such as Rebek and Ghadiri have in their laboratories. At present, most people would probably not consider a set of computer instructions to be "alive." However, these sets of instructions certainly exhibit some very lifelike forms of behavior.

The problem is likely to become a much thornier one if Ray or other scientists succeed in causing artificial multicellular organisms to evolve a great deal. Some of these forms would likely become very complex indeed. Suppose some of them developed intelligence? At first this may sound like a ludicrous idea; however, it is one that many scientists take very seriously. During an artificial life conference that was held in 1990, scientists spent a session debating such questions as to whether artificial life forms should one day be granted civil rights. In his book, *The Dreams of Reason*, physicist Heinz Pagels asked, "The day will come when people will have moral concerns regarding artificial life—what are our obligations to the beings we create?"

Pagels' book appeared in 1988, before the work I have been describing began. He was thinking more of research in artificial intelligence than of what I have been calling artificial life. At the time, nothing approaching the complexity of Tierra had been developed. However, because evolution can take place within a computer, and because massive computing power can allow it to take place at a rate that may be millions of times faster than that at which it proceeds in the real world, we are likely to see some startling developments. We may indeed eventually find ourselves confronted with such questions sooner than we think.

It is easy to laugh at such ideas, perhaps too easy. Perhaps we should remember that we evolved from entities that may have been even simpler than Ray's electronic organisms: small sets of self-replicating chemicals. We could very well find ourselves looking at artificial life that is evolving in ways that no one had imagined. In an artificial world, evolution can literally happen overnight. When evolution proceeds so rapidly, it is hard to say where it might go.

6

IS NATURAL SELECTION
THE WHOLE STORY?

There exist two basic viewpoints about the origin of life and the evolution of living organisms. One says it is not necessary to consider anything other than natural selection. Selection acted on self-replicating chemicals that preceded the beginning of life. The chemicals that succeeded in copying themselves most efficiently left the most "offspring" and soon became common in the primordial soup. Life may have begun either with RNA or with proteins. Whichever was the case, they were chemicals which performed the task of replication better than others. Once life did begin, it was the organisms that were best adapted to their environment that survived and produced the greatest number of offspring. As time went on, living organisms became progressively more diverse and complex, finally producing the biological world that we see around us today.

This is the view expressed in such books as Richard Dawkins' *Climbing Mount Improbable*. Dawkins argues that natural selection, acting on chance mutations, is all that is needed to create the most improbable biological structures (as well as those that surprise us somewhat less). Dawkins discusses such subjects as the evolution of the eye, and evolution of the fig, which is pollinated by a female wasp who pollinates the flowers inside the fig and lays her eggs in some of them. She then promptly dies. The job of reproduction has been done, and her genes will live on in the next generation.

Another point of view is that natural selection cannot possibly explain everything in the biological world. It is filled with complex systems, ranging from gene networks to living organisms and societies of organisms. Complex systems have emergent properties. Thus if we are to really understand the workings of evolution and the nature of life, we must try and find out what properties are created, not by natural selection but by complexity. Natural selection is indeed important, but it isn't everything.

Naturally one doesn't have to adhere to either view in its pure form. Indeed, there are numerous variations on the basic ideas. We have already encountered an example of one of them. Eldredge and Gould do not deny the great importance of natural selection, but they do argue that theories which consider only the action of selection of individual organisms are inadequate. Species sorting, for example, plays an important role. Gould has also argued that spandrels are often encountered in nature. Organisms sometimes have biological structures or modes of behavior that have little or nothing to do with adaptation to their environments. Though he doesn't make this point, spandrels have nothing to do with self-organization either. They are simply byproducts of useful adaptations.

Although I will comment on the controversy throughout this chapter, I will not make any attempt to settle it. That can be accomplished through further research and the testing of new ideas. After all, that is what science is about. It has progressed as much as it has in our century because scientists have expended so much effort in looking for new explanations for natural phenomena, and in performing experiments designed to test those ideas.

My goals are more modest: to give the reader a deeper understanding of what the arguments are all about, and show them where research might lead us in the years to come.

Cosmic Evolution and Biological Evolution

I propose to begin by engaging in a discussion of a topic which is, for the most part, unrelated to problems concerning the nature of evolution and the nature of life. I will give a brief outline of the evolution of the universe. The reader will be able to see how complexity gives rise to emergent properties in the nonbiological world. I hope the examples supplied will also make that rather abstract conception, "self-organizing properties," seem a little more complete.

I will *not* be arguing that if we see self-organizing behavior in the physical world, we should expect to find it in the biological realm also. Arguments by analogy are often treacherous things. To use them here would be even more dangerous than it is in most cases. It is true that the universe is complex, and it has certainly evolved, but there are enormous differences between cosmic and biological evolution. Herbert Spencer, for one, thought that the same principles were operative in both areas. He believed that evolution was an all-embracing cosmic principle. Today it is necessary to be more skeptical. Trying to "read the book of nature" is a much more productive activity than trying to impose our conceptions upon it. Finding emergent order in the universe only suggests that it might be worthwhile to look for it in evolution. Only after we have looked, and looked deeply, will it be possible to draw any conclusions.

A Short History of the Universe

The question of precisely how the universe was created is just as much a matter of controversy as that of how life began. The Big Bang theory, used here, is widely known. Several hundred thousand years after the big bang in which our universe was created, all of space was filled with clouds of hydrogen and helium gas. Small quantities of a few light elements such as lithium also existed. But these were not to play any significant role in the evolution of the universe. There simply wasn't enough of them.

At that time, there existed no carbon, nitrogen, oxygen, iron, or any of the elements with which we are so familiar. Conditions in the big bang fireball had been such that only the lightest elements could be created. Billions of years were to pass before heavier elements could exist in significant quantities. This, incidentally, is not speculation. When astronomers look billions of light years out into space, they are also looking back in time. It follows from the definition of the term "light year" that the light from an object that is a billion light years away has taken a billion years to reach us. Furthermore, analysis of the light from the object enables scientists to determine its chemical composition. The quantities of hydrogen, helium, and other light elements existing in our universe, both in the present era and in previous ones, have been measured.

In some places, the concentrations of hydrogen and helium were slightly greater than they were in other areas. Slowly, these slightly more dense gas clouds began to contract under the force of gravity. It was a slow process; if any observers had been present at this time, it would have been hard for them to tell anything was happening. The contraction was a process that took place over periods of tens or hundreds of millions of years. Approximately a billion years passed between the big bang and the time that the first galaxies began to form.

Gravity was inexorable. As the clouds became denser, the condensation proceeded ever more rapidly. The denser the cloud, the stronger the grip that gravity could maintain on the hydrogen and helium atoms within it. Two or three billion years after the big bang, galaxies had formed, and stars lit up the sky. It must have been quite a sight. But of course there was no one there to see it.

Astronomers are not quite sure which came first: the earliest stars or galaxies. It is possible that galaxy-sized gas clouds formed and that stars were created within them. Alternatively, it is possible that stars were the first to be created, and that gravity then gathered them together in the vast conglomerations we call galaxies. In what follows, I am going to follow the galaxy-first model. If we eventually discover this is not the way things really happened, that the stars were created first, my account won't be altered very much. The basic physical processes are the same in either case.

As time passed, great clouds of gas condensed into galaxy-sized objects. But the condensation eventually had to come to a halt. Practically all astronomical bodies rotate, and galaxies are no exception. There inevitably came a point where the contracting force of gravity was balanced by the centrifugal forces caused by the rotation.

At this point, a process began that was similar to that of galaxy formation. The gas in some regions was denser than that in others.

Thus gravity was able to go to work again and create stars. This time there was a difference. As the protostars contracted, the pressure in their cores increased, producing large quantities of heat. Eventually temperatures in the cores of the stars became high enough, and energies became so great that nuclear reactions began to take place. Some of the hydrogen contained in the stars was converted into helium in a process similar to that in a hydrogen bomb explosion, except that the nuclear "burning" in stars proceeded slowly and evenly. The energy created was converted into light. One by one, the stars began to blink into existence.

At this time, there were numerous stars, but relatively few planets. Most likely, some of the "stars" that were formed were not massive enough to induce fusion reactions within their cores. They became objects somewhat like the planet Jupiter: massive bodies composed primarily of hydrogen and helium. Jupiter can be considered to be such a "failed star." If it were about ten times more massive, ours would be a double star system.

When the hydrogen that fuels an average-sized star runs out, the star gradually dies. Through a complicated process, it evolves into a red giant and then contracts into a white dwarf—a tiny star which continues to glow only because it possesses a certain amount of residual heat. During the death throes of a normal star, some carbon is ejected into space. Thus if only "normal-sized" stars existed, the universe would contain little but hydrogen, helium, and carbon. Obviously, something else had to be going on or we would not be here to talk about it. One could not build a biological world from those three elements. Helium is chemically inert, and carbon and hydrogen can form only a limited number of simple compounds, such as methane. The materials necessary to make substances of the complexity of a strand of DNA, RNA, or protein, would not exist.

Something else was going on. Just as some of the condensing gas clouds were too small to form stars, others were very, very massive. The super-massive stars created had very short lifetimes. They had life spans measured in hundreds of millions of years (by comparison, our sun, an average-sized star, is about 5 billion years old; scientists believe it will live about 5 billion years longer). Not only did the super-massive stars burn out quickly, but their deaths were quite violent. When they died, they exploded as supernovas, expelling elements that had been created within their cores into space.

The pressure within the cores of these stars had been very great, thus the nuclear reactions within them produced a great variety of chemical elements, including many that were essential to life. *All* of the elements heavier than carbon were created in super-massive stars. The requisite conditions simply did not exist elsewhere.

As time passed, second-generation stars began to form. Unlike the first-generation stars, which were made up of hydrogen and helium, these incorporated the chemical elements that had been spread through space by supernovas. More significantly, rings of gas and dust began to form around these stars. When, I use the word "dust," naturally I do not mean anything resembling household dust. On the contrary, this dust was composed of particles of heavy elements.

Gravity went to work once more. The rings began to be drawn together into planets. Some of these were gas giants like Jupiter and Saturn, others were "terrestrial" planets like Earth, Mercury, Venus, and Mars. At first, the surfaces of the terrestrial planets were so hot their crusts were molten. The heat was caused partly by the energy released during gravitational contraction and partly by meteoric impacts.

After about a half-billion years, the planets cooled. The crusts solidified and—at least on Earth—oceans formed. An environment hospitable to life had been created.

What about Biological Evolution?

The story of cosmic evolution is a tale of the spontaneous creation of order. Galaxies formed from hydrogen and helium gas, but not because any external forces acted upon the gas. The protogalaxies exhibited self-organizing properties. The birth of stars was a similar process that took place on a somewhat smaller scale. The creation of the chemical elements necessary to life provides yet another example of emergence. With so much order in the universe, at one time it was believed that this required the existence of a creator. We now realize that, whether there is such a deity or not (this is not a question for science to answer; science, after all, deals with the natural world, not with the supernatural), the fact that complex systems exhibit self-organizing properties is sufficient to explain the existence of galaxies, stars, and planets which provide hospitable environments for life.

But what about biology? Can we automatically assume that there have been times when order spontaneously arose in similar ways? Probably not. There is a fundamental difference between cosmic and biological evolution. When the stars and galaxies formed, they were subject to no external forces. The only significant force that acted upon them was their own internal gravity. Yet in evolution, there is a very significant external "force"—natural selection. It is so important a force that it is entirely possible, as authors such as Dawkins and Dennett claim, that we need consider no other possibilities.

Nevertheless, observing so much self-organization in cosmology should at least give us pause. Is it really plausible to contend that the natural and biological words behave in such different ways? Some scientists, such as the British biologist Brian Goodwin (about whom I'll have more to say later), think not. Goodwin makes no appeal to cosmology, but he does believe that some of the traits that we observe in biological organisms cannot be due to natural selection.

As we have seen, scientists such as Stuart Kauffman and Reza Ghadiri equate the emergence of life with the spontaneous appearance of order. If self-organizing principles were at work during the creation of life, it does not necessarily follow that evolutionary processes must also exhibit emergent properties. It is perfectly possible that self-organizing principles were at work when life was created, and that once the first living organisms appeared, they relinquished their role to natural selection. In this case also, one must be careful not to misled by analogies. One could just as well say that, since no external forces influenced the formation of the stars and galaxies, the same must be true of the creation of planetary systems. Such a conclusion would be totally false. Planets do not form in empty space; the gravitational influences of the stars around which the planets revolve play an important role.

Back to Biology Again

Returning to the theoretical research performed by Stuart Kauffman, I want to look at his work on mathematical models of gene networks. Kauffman claims his models show that it is reasonable to believe that self-organizing principles can be seen in such networks. This is quite an important idea. If Kauffman is right that self-organization exists on the genetic level, then there is every reason to look for it in evolution, and in the biological world in general.

Kauffman uses a mathematical model in which the interaction of genes is represented by mathematical entities known as Boolean networks (named after the nineteenth-century mathematician George Boole). In Kauffman's model, each "gene" has a fixed number of inputs from other genes. These inputs have the function of turning the gene "on" or "off." Kauffman doesn't call them "genes." He refers only to elements in Boolean networks because he wants to emphasize the fact that he is dealing with a very abstract model of biological reality. The network is "Boolean" because Boolean algebra is a system of mathematical logic that represents relationships between objects.

It is not necessary, however, to know anything of Boole's algebra in order to understand what Kauffman has done. The essential point to know is that he needed a way to model relationships between genes mathematically if he was to analyze gene networks on a computer. In the real biological world, a gene that is "on" synthesizes a certain protein while one that is "off" does nothing. So Kauffman's model is an accurate representation of biological reality—up to a point.

Kauffman's theory says nothing about protein synthesis. He is concerned with studying abstract networks so that he might discover what the optimal arrangements are. Furthermore, his networks are more simplified than those that occur in the real biological world. Genes do much more than turn one another "on" and "off," and some of their interactions are rather indirect. For example, one gene may produce protein A and another gene protein B. B might then act as a catalyst in a reaction that transforms A into protein C. In such as case, the two genes must be said to interact even though they don't directly influence one other.

When I point out that Kauffman has not included everything, I am not accusing him of oversimplifying. What he does is simply good science. Centuries ago, scientists discovered the best way to analyze a complicated system was first to examine the simplest cases. Once the basic features of the system being studied were understood, the refinements could be added in one by one. When Isaac Newton wanted to determine the shape of Earth's orbit, he did not begin by trying to compute the gravitational influences of all the bodies in the solar system. He assumed, instead, that these forces were so small that they had only insignificant effects. Considering only the gravitational force exerted by the sun on the earth, he found that Earth would trace out an elliptical orbit. Later on, scientists who wanted to calculate the earth's motion with even greater accuracy added in the gravitational influences of such bodies as Venus, Mars, and Jupiter. Similarly, a physics graduate student who is studying quantum mechanics is taught to understand the simplest atom, hydrogen, first. This gives him a basic understanding of the behavior of atoms. He can then go on to more complicated cases.

Kauffman has been studying Boolean networks for many years. He began by doing pencil and paper calculations on networks with small numbers of elements. He then went on to study more complicated ones on computers. His first computer calculations were done in the days when the machines were relatively slow and still used IBM punch cards. At one point he astonished his colleagues by shuffling a stack of IBM cards before returning them to the system. In those days, if even one card was out of place, the program they contained would not run. Of course, Kauffman wasn't quite so mad as he appeared to be. He wanted to introduce an element of randomness into his networks. Today much faster and more sophisticated machines are readily available to anyone who wants to use one, and such tricks are not necessary.

When Kauffman looked at different kinds of networks, he achieved some very interesting results. He found, first, that if each element of a network had just one input from other network elements, then the system showed a tendency to freeze. Genes would tend to fall into states where they were permanently on or permanently off. Obviously, this could not correspond to anything happening within a real cell. The system was just too stable; it was as though there was no interaction between the genes at all. This was probably a good model of a cell that had died, but not of anything that went on inside a living organism.

Yet, if the number of inputs per gene was three, four, or five, or greater, the behavior of the system became chaotic. The system would pass through an enormous number of different states without ever repeating itself. Here, "state" means any configuration in which certain genes are on and others are off. Suppose one has a small collection of genes labeled A, B, and C. The configuration in which A is on, and B and C are off is one state. A off, and B and C on is another.

Clearly, this could not be a good representation of gene activity within a cell either. If the genes behaved in a chaotic manner, and were doing something different at each moment, there would be little or no coordination between them. They could not possibly regulate cellular activity. In such a situation, cells would most likely not exist.

Finally, there was the case in which each network element had two inputs. Kauffman found that the genes would then cycle through a relatively limited number of different configurations. When the cycle ended, it would be repeated all over again. Here was a kind of orderly activity that could serve as a model for genetic activity in real cells.

When I tell you that, after years of work, Kauffman found Boolean networks with two inputs per element had certain properties, you may not be very impressed, but you should be. Kauffman had discovered that, out of the numerous different states certain types of complex systems could have, there was one that possessed very special properties. For the first time, someone had discovered a law that governed the behavior of complex systems.

The number "two" in two inputs per element has little significance. If one introduces further refinements and looks at networks that are more complex (a number of different scientists have done or are doing this), the actual number can vary. The significance of Kauffman's discovery was that certain complex systems had special states. They were located outside the "frozen" regime where nothing much happened. But they were just far enough from the chaotic region that they were able to exhibit emergent properties.

In Kaufman's terminology, the two-input case lies at the "edge of chaos." Kauffman did not invent this term. It was coined by the computer scientists Chris Langton and John Holland. It is Kauffman who has made the most extensive use of the idea. He speculates that all living systems seek out the edge of chaos. It is this, he thinks, that makes life and evolution possible.

Initially, Kauffman thought that genes in a living cell would automatically seek out such a configuration without the intervention of natural selection. When Kauffman's "edge of chaos" idea first came to him, he found it so stunning that he began to doubt whether natural selection played any role in evolution at all. Perhaps self-organization could account for everything? If genes could organize themselves in such a way, couldn't this be the most significant factor in the evolution of life?

Nowadays, Kauffman expresses a more conservative viewpoint. Life exists at the edge of chaos, he states, "because evolution takes it there." In saying this, he is not denying that gene networks have emergent properties. The fact that there is an optimal configuration puts severe constraints on natural selection. It *must* bring the life to the "edge of chaos" if it is to produce viable organisms capable of evolving further.

In a way, the change in Kauffman's outlook is similar to that of the geneticists of the early twentieth century. At first, they thought they had done away with the idea of natural selection, that naturally occurring mutations were all that was needed if evolution was to take place. Later, they reversed their position when it became apparent that mathematical genetics could explain all the details of natural selection. Genetics showed precisely how mutations would spread throughout a population.

The analogy isn't exact. Though Kauffman no longer believes that he has found an idea which will replace that of natural selection, he still maintains that natural selection isn't everything. The fact that complex systems have emergent properties plays a role in evolution, too.

But What Does It Have to Do with Reality?

Real species evolve by adapting themselves to their environments. In each generation, the less well-adapted will die or fail to reproduce, while the better adapted will pass on their genes to the next generation. To be sure, there is some evidence that, in many cases, selection does little but keep organisms pretty much the same if they live in a relatively stable environment. Something like this must be the case if Gould's and Eldredge's ideas about punctuated equilibrium are correct. Environments are not always stable, and few doubt that adaptation by natural selection is what causes evolution to take place.

It would be interesting, therefore, to see whether Kauffman's Boolean networks could evolve in such a manner. Unfortunately, it is not so easy to see how this could be accomplished. His networks have no environments; they are simply numbers which exist inside electronic computers.

However, Kauffman has found a way around this. He has shown that his networks can coevolve. Coevolution is a concept that is based on the observation that organisms do not exist in isolation. They populate environments that are also occupied by other living beings. Thus, more often than not, they find themselves cooperating or competing with other species. Two species may compete for the same food resource, for example.

Another kind of coevolution is seen in predator–prey relationships. If antelopes learn to run more quickly, the lions must evolve in turn if they are not to go hungry. Perhaps they will learn to run faster too, or evolve more powerful leaps, enabling them to spring upon their prey more quickly. If frogs develop sticky tongues, then the flies will have to evolve some defense (perhaps slippery feet) if their population is not to be decimated.

University of Chicago paleontologist Leigh Van Valen has named this the Red Queen effect. Just as Alice and the Red Queen in Lewis

Carroll's *Through the Looking Glass* found themselves having to run very hard just to stay on one place, species that are involved in predator–prey relationships often find themselves having to evolve quickly, Van Valen says.

In some cases the relationships are cooperative rather than predatory. Flowers and the insects that pollinate them evolve together. Presumably, so did human beings and the *E. coli* bacteria which live in our intestines. We provide an environment in which these bacteria can thrive, and they help us to digest our food.

Now Kauffman could not set up predator–prey relationships between his Boolean networks, or even have them compete for food (after all, they don't eat). Nor was there any obvious way to make them form symbiotic relationships. But he could have his networks interact with one another by matching one another's on–off patterns. When random elements analogous to mutations were introduced into the networks, evolution did indeed take place.

Kauffman has also done mathematical analyses of the relationships between his networks and fitness landscapes. It was this research that allowed him to draw conclusions about the relative merits of random, smooth, and "rugged" landscapes. If Kauffman's models bear any relation to reality, it appears, they are capable of telling us a great deal about evolution.

These results are open to the same criticisms that have been leveled against his work on autocatalytic sets. Because they are based on mathematical models, it is not clear what his findings imply about the behavior of genes, cells, or organisms in the real world. To be sure, Kauffman's ideas are original and provocative. Until some way is found to test them experimentally, there is no way of knowing whether or not they are true.

Speaking about computer simulations in general, Manfred Eigen has put this idea succinctly. "A theory has only the alternative of being right or wrong," he says. "A model has a third possibility: it may be right, but irrelevant." Similar ideas have been expressed by the Oxford University ecologist Robert May. Commenting on the work done by Stuart Kauffman and other scientists at the Santa Fe Institute, May said the computer models that have been developed at the Institute are "mathematically interesting but biologically trivial." Other biologists have echoed the thoughts of Eigen and May by expressing the opinion that the bulk of the work done at the Institute is exclusively mathematical and has little bearing on real biology.

One shouldn't take such pronouncements as gospel. Much of the work done at the Institute is quite original. The scientists there are seeing approaches to old problems. Original, new ideas often encounter a great deal of opposition. When Einstein published his special theory of relativity in 1905, many scientists concluded that this new theory didn't look like any physics *they* knew. When Einstein received the Nobel Prize in 1921, it was for his contributions to quantum theory, not for relativity. There were still too many doubts in the minds of the members of the Swedish Academy about the latter. Indeed, it has been said a new theory that breaks new ground generally becomes established, not because its opponents are convinced of its validity, but because they eventually grow old and die.

However, originality hardly guarantees the truth of a new idea. Eventually, a theory must be confirmed by empirical evidence if it is not to wind up on the trash heap. As the nineteenth-century British biologist T. H. Huxley once remarked, many beautiful theories had run afoul of "ugly facts."

Kauffman is not afraid to extend his speculations to fields outside the discipline of biology. Kauffman has even wondered whether his ideas might not apply to the evolution of the universe itself. I won't go into much detail here, for he stresses that he is dealing only in very tentative ideas, and issues several disclaimers. His ideas, he says, are "not yet science." He has not published these speculations in a book or in a scientific paper, and it is not unlikely that he may decide to modify them in one way or another. Since he may very well have changed some of them by the time this book appears, I suggest that you check out his home page at the Santa Fe Institute's website if you're interested. You'll find them posted there. (If you don't have an online connection, try the nearest library; most of the larger ones are connected to the Internet these days.) If you look up Kauffman's speculations (see "World Wide Web Resources" section at the end of the book for his address) you'll have a chance to see "science in action."

Kauffman possesses an original and far-ranging intellect. This has gained him a great deal of respect within the scientific community. A number of times, I have stressed that his theories could very well turn out to be wrong. Even if they do, he will have enriched the science of biology. Matters always get livelier when someone has the audacity to question long-established theory, and to suggest ways in which it might be modified or replaced.

A Little Organism with a Big Name

Unlike Kauffman, the British biologist Brian Goodwin has studied the behavior of a real organism. Well, sort of. In order to understand certain aspects of its behavior in nature, he, too, created computer simulations. They dealt not with abstract ideas, but with an organism that could be seen and studied. Goodwin's computer model suggested a possible answer to a question that had been troubling him. The results that he obtained were quite striking.

Here, I find that I must insert another caveat. Computer models can indeed be quite useful. Astrophysicists, for example, use them all the time in an effort to understand such phenomena as the formation of galaxies. But astrophysicists have a great deal of empirical data at their command. This allows them to check the validity of their computer-generated results. Those who want, like Kauffman and Goodwin, to demonstrate the existence of emergent behavior in complex biological systems generally do not. Possibly I have been belaboring this point to the extent that the reader is finding my emphasis (or overemphasis) a little tedious. I make no excuses. In the end, scientific knowledge must be based on facts. That cannot be said too many times.

Acetabularia acetabulum (common name: Mermaid's Cap) is an alga found around the shores of the Mediterranean. In the adult form, the alga has a stalk that is one or two inches long, and has a cap at the end. It is a single-celled organism with a nucleus that is located in one of the branches of a rootlike structure at its base. *Acetabularia* evolved about 600 million years ago. Today, it does not exist in such great abundance as it did in the past. Nevertheless, it should be around for many millions of years to come.

Acetabularia exhibits a peculiar sort of behavior. As it grows into its adult form, rings of tiny leaves, called whorls, form around the stalk. The process is repeated many times. The whorls do not perform any obvious function, and they soon drop off. This raises the question: why should *Acetabularia* invest time and energy in producing structures that are quickly discarded? Why hasn't natural selection eliminated such wasteful behavior? In the human body there are structures, such as the appendix, that serve no obvious purpose. Natural selection hasn't eliminated these. But then, we haven't been around for 600 million years. Evolution may eliminate the appendix in time.

Goodwin modeled *Acetabularia* on a computer to determine why it produced whorls. In particular, he looked at the concentrations of certain chemical elements present when his model organism grew. He found that—on the computer at least—the concentration of calcium increased in certain parts of the alga during growth. When it did, the cell wall softened. The areas of highest calcium concentration resembled rings, and they were found at locations where the whorls formed in the real *Acetabularia*.

Goodwin's computer model was unable to produce whorls. Nor did the structure that formed at the end of the stalk when growth was complete look anything like the real *Acetabularia*'s cap. Nevertheless, Goodwin had achieved a striking result. The computer model suggested that whorls were formed, not because this process had been preserved by natural selection, but because the whorls were part of the natural process of growth. Goodwin had found that, in at least one organism, certain structures were created, not because natural selection had made them, but because certain patterns arose spontaneously.

It is possible to observe other patterns in nature which seem to have no adaptive purpose. For example, the seeds in a pine cone arrange themselves in spirals. Some of these wind up to the left and some to the right. If you look at a cone, you may see eight right-handed spirals and thirteen left-handed ones. Alternatively, the numbers of spirals may be something like thirteen and twenty-one. These numbers are members of the Fibonacci series, named after the thirteenth-century Italian mathematician Leonardo Fibonacci.

In this series, each term is the sum of the previous two; the first few numbers are 1, 1, 2, 3, 5, 8, 13, 21, 34, ... Note that 3 is the sum of 1 and 2, that 13 is the sum of 5 and 8, and so on. All this is very interesting. But why should pine cones make use of such a series of numbers? Fibonacci himself would have been surprised. He devised the series in order to solve a problem involving the breeding of rabbits. The same phenomenon (double spirals using Fibonacci numbers) is seen in other plants. The seeds of a sunflower are arranged in such a manner, for example.

The appearance of patterns such as those seen in pine cones (and sunflowers) is known as phyllotaxis. The existence of this phenomenon is a mystery. Possibly natural selection produced it in some way that we cannot fathom. But why? How? Could it be that it is an example of the spontaneous appearance of order in nature? This is indeed possible, but again we are at a loss to explain how it arose.

The existence of phyllotaxis hardly confirms Kauffman's and Goodwin's ideas about the spontaneous appearance of order in nature. It is difficult to find many other examples, while the number of biological traits that were apparently molded by natural selection is legion. The swiftness of the antelope must be ascribed to selection, as must the stickiness of a frog's tongue, and the length of the giraffe's neck. Such examples could be multiplied endlessly.

It appears we must conclude that, like Kauffman, Goodwin has some original, provocative ideas that are far from being confirmed. It doesn't necessarily follow they are wrong. Too little empirical observations have been made to allow us to decide. Evolutionary biologists have not looked for examples of emergent order in nature. Instead, they have focused on explaining natural selection on the level of genes. Kauffman and Goodwin and their colleagues may eventually be proved wrong, but they have opened up a debate that could turn out to be very fruitful.

Philosophical Differences

According to Stephen Jay Gould, chance and contingency play an important role in evolution. If the Cambrian explosion could somehow be run again, he says, most likely entirely different sorts of creatures would evolve. As a result, the biological forms that existed today would be very much unlike those we see around us. Nothing resembling human beings would exist.

Brian Goodwin doesn't believe it. In his view, only a limited number of biological forms are possible. There exist laws (yet to be discovered) of form and structure that constrain how biological organisms are constructed. In other words, the main features of biological design are expressions of certain natural patterns. We may not know what these patterns are, but there is evidence that they exist. Goodwin believes that Gould is wrong, and thinks if the Cambrian explosion were run again pretty much the same kinds of organisms would evolve. Goodwin doesn't deny that natural selection is operative in the biological world. He refuses to admit that it is all important. Goodman says that he would give natural selection "a 5 on a scale of 1–10."

Though Gould and Goodwin hold opposing views, they haven't debated their differences, either at scientific conferences or in print. Gould has had some kind things to say about Goodwin. He links Goodwin to "one of these great traditions of Western thought, structuralism in biology." In pre-Darwinist days, biologists were very concerned about questions concerning the form and structure of biological organisms. In Gould's view, Goodwin has brought some of these ideas back.

Daniel Dennett take a very different point of view. He has stated flatly that he thinks Goodwin's ideas are "mainly wrong." He characterizes Goodwin as a "romantic, who wants to deny that biology is ultimately engineering." He goes on to say that, "all the regularities of biology strike me as being exactly like the regularities of engineering," and comments that Goodwin is making "an implausible claim."

When Dennett compares biology to engineering, he is not, of course claiming that organisms had intelligent designers. He simply sees similarities between intelligent design and the designs created by natural selection. "Locomoting organisms tend to have the eyes and mouth at the front end," he says. "It's not a deep law of nature; that's the way to design something that has to fend for itself."

"Goodwin is unhappy with that style of explanation," he adds, "and would like to see deeper laws of physics explaining all this. It's an idea I think I do understand—and don't believe at all."

The comments about Goodwin's work are an excellent illustration of the rift that exists between the complexity advocates and traditional evolutionists. There are those who, like Gould, are open to new ideas, but who nevertheless remain committed to the view that natural selection is the primary operative force in evolution. And there are those who, like Dennett, consider natural selection to be the only operative force.

In a way, this is like listening to an argument between an atheist and a theist. The atheist "knows" that God doesn't exist, and the theist "knows" that God does. Neither can come up with much evidence to support his point of view so the argument is likely to go on endlessly. The debate that we are currently witnessing may not last that long. It's a pretty good bet that it will until more empirical evidence is available. This is not evidence that is likely to be found tomorrow. At the moment, no one really knows where to go to look for it.

The reactions to Kauffman's work have been more favorable than those to Goodwin's. But Dennett, for one, has written favorably of Kauffman. In *Darwin's Dangerous Idea*, he calls Kauffman a "meta-engineer," and places him squarely within the Darwinian tradition. According to Dennett, Kauffman has done biology a great service by pointing out the significance of the "evolution of evolvability."

More often than not, one encounters two opposing philosophical outlooks. I say "philosophical " and not "scientific" because, at the moment, the evidence that would cause us to either accept or reject the ideas of the advocates of complexity is lacking. When scientific evidence does not exist, one must rely on philosophical "gut" feelings instead. I would bet that Goodwin "knows" there exists only a certain number of natural biological forms, while Dennett "knows" that biological organisms are only (in the words of the French biologist François Jacob) "tinkered-together contraptions."

7

ARTIFICIAL LIFE
ON THE INTERNET

E volution has happened only once on earth. We are all descendants of organisms, probably a single organism, that evolved about 3.5 billion years ago. DNA is the same in all living beings, plants and animals alike; the same DNA sequences code for the same proteins. This is evidence for a single line of descent. The earth has witnessed one Cambrian explosion. We and all other multicellular life-forms on earth are descended from the organisms that proliferated at that time. It is true that there was probably multicellular life before the Cambrian period. Either all the creatures that existed died out or some of them were ancestors of the life that proliferated during the Cambrian era.

Having only one example of evolution to study places certain limits on our understanding. If scientists could only study evolution on other planets, they could get an idea as to what features of evolution always occur and which are unique to life on earth. For example, was the Cambrian explosion—or something like it—inevitable? For that matter, must multicellular life evolve in the first place? Is it something that is very common in the universe? Or is it very rare? Is punctuated equilibrium an essential feature of evolution? Or could a gentle gradualism sometimes be the rule? Discovering other kinds of life would answer some of these questions. It might also answer the question of whether a genetic code has to be carried by something resembling DNA or whether something very different is possible. Discovering extraterrestrial life would likely give us insights the nature of which we can only guess.

If we knew the answers to such questions, this would deepen our understanding of the workings of natural selection, and we might find a definitive answer to the question of whether natural selection is the sole operative factor in evolution. If we could study extraterrestrial life, scientists might not have to argue endlessly about such matters. They could appeal to the accumulated data instead.

Tierra

These and other considerations have given scientists the idea of trying to see if they could study evolution inside computers. If electronic organisms were created that had the ability to reproduce and evolve, we would have the opportunity to look at evolution in another setting. We would then have, not just one example of life, but two. This idea occurred to ecologist Tom Ray during the late 1970s. He was not the only person to whom it occurred. However, he has done the most with it.

Ray was an ecologist who had little experience with computers. So at first he was unable to pursue the idea. Though he had thought of the idea of electronic evolution during the late 1970s, it was only in 1989, when he became an assistant professor at the University of Delaware, and was able to make use of the computing resources available at the University, that he was able to pursue it. At this time, Ray dedicated himself to learning about computer programming and technology. Before long, he knew quite a bit about these subjects— for a biologist.

I have already related the story of Tierra in Chapter 5. However, it wouldn't hurt to review some of the details at this point, while elaborating a bit on some of the things I said earlier. This will make my account of Ray's next project, the successor to Tierra, a bit easier to understand. When Ray started the experiment, he seeded his artificial world with an electronic organism that he called the Ancestor. It lived in the computer's RAM memory and used the CPU (central processing unit) as a source of energy. It was subject to mutation in several different ways. First, random mutations were possible; these were analogous to mutations caused by cosmic rays in biological organisms. Mutations also sometimes took place when the creatures replicated themselves; this had the effect of introducing genetic variation somewhat like that which is created when biological creatures mate. In biology, the organism that is the product of the mating inherits half its genes from each parent; many new combinations of genes are possible. A third type of mutation was somewhat more complicated. The computer instructions within a creature were sometimes altered when they executed their code. One could say that being "alive" and interacting with the environment caused a mutation from time to time.

Of course there was the reaper—the subprogram that killed off older or less well-adapted organisms and prevented Ray's electronic world from becoming overpopulated. A very well-adapted organism might live longer than average, but it could never escape the reaper entirely.

In nature, parasites tend to be less complicated organisms than their hosts. They are also more numerous. There are many more different species of parasites in our world than there are of hosts. I suppose that this statement might cause you to raise an eyebrow. But I think you will become convinced once you begin to consider the numerous different kinds of infections and infestations that humans can get. Viruses and fungi are parasites too. The class isn't limited to things like tapeworms.

Evolution typically does away with the functions that parasites do not need. A parasite that attaches itself to a particular part of a host's body has no need for locomotion, for example. A virus doesn't even need a metabolism; it makes use of the metabolic machinery of the cells that it infects. Ray found that, in his simulation, parasites also gave up unneeded functions. Although the ancestor had a set of eighty computer instructions, the first parasites that appeared had only forty-five. They never would have been able to live and reproduce on their own. But once they gained the opportunity to make use of parts of the hosts' metabolism, they didn't need the extra code. With forty-five instructions, they did just fine.

The host–parasite system was found to exhibit behavior similar to that observed in nature. At first, the parasites proliferated and the hosts, which had no defenses, declined in number. This led to an overabundance of parasites, which then began to die off for lack of "food." Now that the hosts were temporarily free of parasites, they again proliferated. Tierra had produced a cycle of a sort that was well known to ecologists, one that is seen in many settings where there is a predator–prey relationship. At first, the predators will kill off their prey, and the predator population will decrease. This causes a rise in the prey population, which will be followed by a subsequent increase in the number or predators. The cycles of increasing and decreasing population will be repeated, although never in exactly the same way. For example, there may be some occasions when either predators or prey may almost die out (if the populations are small enough, chance fluctuations can wipe out one population or the other, causing it to become extinct). They will then increase their numbers again. The cycles are quite well known, and can be described mathematically by what ecologists call the Volterra–Lotka equations.

If Tierra was allowed to run still longer, a kind of evolutionary arms race ensued. The hosts evolved into organisms that could attack the parasites. They soon drove the latter to extinction. Ray called these "hyper-parasites"; they were genetically different from the ancestor by a factor of about twenty-five percent. After hyper-parasite evolved, Tierra had a tendency to go into a period of temporary stability. But then new kinds of parasites appeared which were able to take advantage of certain genetic characteristics that the hyper-parasites possessed. It seems that the hyper-parasites quickly learned to cooperate symbiotically. The new parasite would place itself between two cooperating hyper-parasites and steal pieces of the hyper-parasites' computer code. This new organism, a kind of hyper-hyper-parasite, was only twenty-seven computer instructions long.

If you use a computer, you can observe electronic evolution yourself if you like. Tierra can be downloaded free from Tom Ray's Tierra home page. It will run on DOS, Windows, Macintosh, Unix, VMS, and Amiga platforms. Be aware that Tierra is a scientific research tool, and that it is more difficult to set up and use than commercial software. If you decide to try running Tierra, it would be helpful to have some computer expertise. In any case, for more information, see the "World Wide Web Resources" section after the bibliography at the end of this book for software resources.

And What Does All This Have to Do with Evolution?

The development of Tierra shed little or no light on the dynamics of biological evolution. It was just too simple. For example, the eighty computer instructions of the ancestor were arranged into just three genes, making this organism hundreds of times simpler than the least complex bacteria found in nature. Furthermore, some aspects of Tierra are hard to interpret. The results that were obtained sometimes varied depending upon the computer language that was used. The digital evolution in Tierra may proceed slowly and gradually with one language and show periods of punctuated equilibrium when used with another. It is hard to guess what this means because there is no theory which relates the structure of a computer language to the patterns of electronic evolution. It is true that the punctuated equilibrium observed in some systems is significant. But one would like to know exactly what it is that makes evolution (either real or digital) proceed in this manner.

Tom Ray's great achievement was to show that electronic evolution was possible, and his success with Tierra suggested what the next step should be. If he could create digital life of greater initial complexity and place it in a more complex environment, the potential for evolution would be greatly enhanced. Ray has developed a method for accomplishing this. His experiment, which involves setting up a Tierra-like environment on the Internet, has already begun as I write this and will still be going on when this book is published.

Stephen Jay Gould has argued—rightly, I think—that life on earth is and always has been dominated by bacteria, and that most living organisms are relatively simple. However, to us, evolution really only becomes interesting when we look at the history of life following the Cambrian explosion. It was then that new levels of complexity began to evolve. The explosion of diversification which occurred then arouses our curiosity more than the history of single-celled organisms. This is more than just anthropomorphism (or mammal-centralism, if you like). The life-forms which arose during the Cambrian period developed nervous systems, immune systems, ecologies, new means of locomotion, and eyes, to mention just a few of their characteristics. They could do many things that simpler creatures could not do. Eventually (though perhaps not inevitably), this lead to the evolution of intelligence.

How is one to bring about the evolution of such complexity in a computer? The RAM memory of a single machine does not provide a rich enough, or a spacious enough environment. For example, the computer on which I am writing (which has become somewhat obsolete two and a half years after I purchased it) contains 16 megabytes (16 million bytes) of RAM memory. Compared to numbers which describe the complexity of living organisms, 16 million is a very small number. To cite just one example, the human cortex contains about 100 billion neurons. It was considerations of this sort that caused Ray to realize it was unlikely that very complex electronic life could evolve in the environment provided by a single computer, even a supercomputer. Naturally one could begin the experiment with a more complex ancestor. But this would probably only produce a Tierra-like evolution all over again.

It wasn't difficult to see that there was a way around this problem. Computers are interlinked via the Internet these days. Those of us who use Macs or PCs are connected only when we log on. But mainframe computers used at scientific computing centers generally maintain their Internet connections all the time.

So why not create electronic creatures that could migrate around the Internet? Their environment would certainly be much more diverse than that provided by a single computer. If CPU energy was available to them only when a computer was not being used for some other purpose, they would be induced to move from site to site. Very likely, different digital organisms would become adapted to different environments. Electronic life would become more diverse. There was every reason to hope that complex electronic ecologies would develop. If their evolution were allowed to go on long enough, Ray hoped, multicellular electronic life would become complex enough that something resembling the Cambrian explosion would ensue.

As I write this, trial runs involving hundreds of computers have alrcady been made, and Ray and his colleagues are working to eliminate bugs from the program. As the new version of Tierra is perfected, more and more computers are likely to be linked up. Ray's organisms only feed on CPU energy when a computer is not being used for some other purpose (at night, for example), so acting as a host to Ray's creatures does not impair a machine's computing abilities. The fact that CPU energy is only available to Ray's electronic organisms some of the time is an asset, not a disadvantage. It creates additional environmental features to which they must adapt.

Sex

No one really knows why sex exists; it is one of the great puzzles of evolutionary biology. After all, the organism that is the most fit is one that passes on the greatest number of its genes to the next generation. If it reproduces asexually, all of its genes are found in each of its offspring. In sexual reproduction, only half are passed along. The offspring have only half of the genes found in each parent. Why should natural selection favor such an arrangement?

It does no good to argue that sexual reproduction increases genetic diversity, thus making a species more adaptable and able to adjust more rapidly to changing environmental conditions. Natural selection knows nothing of the good of a species. It works at the level of the individual. The fittest *individuals* pass on their genes. The less fit do not. Biologists don't really know how to answer the questions raised by the existence of sex. There have been a number of hypotheses, but none has gained wide acceptance. We know that sex is a desirable thing, but its origin remains a mystery.

From a Darwinian point of view, sex has a number of disadvantages, at least from the standpoint of the individual who engages in it. The fact that an organism passes only 50 percent of its genes along to the next generation, rather than 100 percent, has already been cited. In addition, time and resources must be invested in finding and courting a mate, time that otherwise might be spent foraging for food. Sometimes sex can be downright dangerous. An impressive set of tailfeathers may make a peacock appealing to the peahens, but it is also likely to attract predators.

The existence and persistence of sex is a puzzle. However, since sexual reproduction does increase genetic diversity, it was desirable to have it in the network version of Tierra. Ray's intention, after all, was to evolve as much complexity as he could. If he had thought of a way to accomplish it, Ray would have left sexuality under the control of the organisms he was evolving. Unfortunately, there seemed no obvious way to do this, at least not in Tierra. So he introduced "genetic operators" which, in effect, forced his creatures to have sex. In Ray's mind, doing this was a "trivial sin." His creatures would have no control over their sex lives, but they would evolve more rapidly, and would thus be able to do many more things for themselves.

Sex in Tierra is not like mammalian sex. It is part of a two-step process. First, a mother organism will replicate "herself" (I put herself in quotes because there are really no sexes in Tierra), producing a daughter organism with an identical genome. The daughter will then be induced to find a mate with which "she" will exchange genetic information. Reproduction and the exchange of genes are two steps in a two-part process.

This kind of sex may seem bizarre to human beings, but it is really quite similar to forms of sexual activity observed in nature. Bacteria, for example, engage in an activity called conjugation during which they will exchange small pieces of genetic material with one other. This activity has nothing to do with reproduction; it is simply a method of spreading genetic diversity through the population. Bacterial sex, incidentally, is one of the causes of the rapid spread of resistance to antibiotics. If one bacterium develops resistance, it can give the appropriate gene or genes to other organisms. Sex in Tierra isn't quite the same, because the amount of genetic information exchanged is greater. For example, in conjugation, the DNA that is exchanged is not nuclear DNA (i.e., DNA found in a cell nucleus), but rather DNA that floats around inside the cell body in small objects called plasmids. But there are similarities.

Tierran sex is odd but efficient. It should be capable of producing as much genetic diversity as the variety that "higher" creatures use.

The only problem that might appear is that, when the Tierran organisms become intelligent, they might begin to object to having sex against their will. When I say that, I am only partly joking. Tom Ray believes that the evolution of intelligent life in a Tierran environment is a real possibility. If he is correct, his creatures would naturally become aware of the mechanics of their brand of sexual reproduction. The evolution of electronic intelligence will be discussed after I describe some more of the details of network Tierra.

Problems, Problems

As I mentioned before, Ray set up network Tierra to investigate the evolution of complexity. Although it might have been interesting to start with single-celled creatures and see if multicellularity would evolve, this would not have been very relevant to the main purpose of the experiment. So the ancestor that is injected into the Tierra environment is multicellular to begin with. To be sure, it is not very complicated. The original Ancestor had ten cells and two cell types. There were eight sensory cells and two reproductive cells. The number of cells in organisms that have evolved from it has not been very much larger or smaller.

The creation of sensory cells (which were necessary if the electronic organisms were to travel from one computer node to another) caused an unforeseen problem. Since sensory cells were not essential to reproduction, and since natural selection favored smaller organisms, which could reproduce faster, the sensory cells tended to be eliminated.

The problem was eventually solved, but it was only one of many that Ray encountered soon after he created the system. The ancestor he seeded into the network Tierra system contained 640 computer instructions. It was much larger than the ancestor used in the original version of Tierra, which contained only eighty. It was only to be expected that if one wanted to let more complicated organisms evolve in a more complex environment, there would be a lot of bugs to be eliminated and a lot of problems to be solved. This indeed turned out to be the case.

Sometimes organisms attempting to migrate from one computer node to another failed to reach their destinations. In one early run, some ninety percent were lost in this manner. Ray found, next, that it sometimes happened that no new organisms were born on some nodes. The inhabitants of such a node would send their offspring to other ones. Thus the creatures that inhabited the "home" node aged, accumulated mutations, and eventually ceased to function. They didn't participate in the evolutionary process at all. And then there were organisms that behaved in the opposite manner. They tended not to migrate, but simply stayed on the same node and reproduce. Naturally, this was also contrary to the purposes of the experiment. Reproduction and evolution on a single node had already been studied in the original version of Tierra.

Meanwhile, some nodes became "tar pits," death traps for any creatures that were unlucky enough to emigrate there. Finally, certain kinds of odd-behaving mutants appeared. For example, one species did nothing but emigrate. As soon as it reached a node, it would immediately execute an instruction that caused it to bounce somewhere else. These "bouncers," as Ray called them, seemed to prefer to live in the spaces between nodes rather than remain for any length of time on any individual computer. Their behavior made a certain kind of sense, because they were thus able to avoid the reaper. However, Ray had set up the experiment to study the evolution of complexity, not to watch his electronic creatures bouncing back and forth across the Internet.

There were also bugs, plain old computer bugs. Every program has them. If Windows 95 crashes a little more often than you'd like, it's probably caused by a bug of one kind or another. If you can't get into a weapons shop while playing your kid's (or your) role-playing game, you can reasonably blame it on a bug. There are really no exceptions. Unix, the operating system commonly used to run mainframe computers, does have relatively few. It has been around for a long time, and there has been ample opportunity to correct the most common problems. That is why Unix is stable when compared to the Windows and Macintosh platforms. I would be very surprised if someone were to discover that Unix has no bugs at all.

The network version of Tierra makes use of newly created software. It is no surprise that bugs have been a problem. As I write this, the system is functioning, if not always in the ways that Ray would like. Though there is still work to be done, the experiment is beginning to produce results. Meanwhile, if you have Internet access (either at home or at the library), you can keep up on the progress of the network Tierra by going to the Tierra home page (see the "World Wide Web Resources" at the end of this book for the Web address). You will find that Tom Ray regularly posts progress reports, which keep the scientific community (and anyone else interested) informed of the latest results, often months before they are published.

The Evolution of Complexity

Ray set up the experiment to study the evolution of electronic organisms in a complex environment. Naturally he doesn't know quite what he will find. It is not unlikely that the experiment will shed some light on the evolution of complexity in the biological world. Biologists do not know why there was a sudden explosion of diversity during the Cambrian period, or why organisms began to evolve that were more complex than those that existed before. Fossil evidence tells us that there was animal life before the Cambrian era. According to developmental biologist Eric Davidson of the California Institute of Technology, life might have spent tens or hundreds of millions of years as simple assemblages of cells. He calls them "squishy little larvalike things." Why, then, did everything change during the Cambrian period? As yet, there are no answers.

It is possible that, as fossil evidence accumulates, we may have to revise some of our ideas about the evolution of life. New discoveries, for example, could show that there was more diversity in the pre-Cambrian period than scientists currently think. This would not alter the nature of the problem, it would only change the dates. Scientists would still have to explain why and how animal diversity and complexity arose. So far, they have not been able to do so.

If a sudden increase in complexity is observed in the creatures that inhabit network Tierra, there will be no guarantee that biological life became complex in the same ways. At the very least, observations of increasing complexity in electronic life might provide biologists with some ideas.

One of the unsolved problems of evolutionary biology has to do with speciation—the creation of new species. Biologists understand why there exist a number of different species of antelope. Geographical isolation of different populations can explain that. All that is required is that, during the period of isolation, one population changes enough so that its members can no longer mate with members of the other and produce fertile offspring. The differences don't have to be very great. To the nonscientist, members of two different species of antelope would probably look very much alike. Yet, they cannot productively mate, and thus they will continue to evolve along separate lines. For that matter, a horse and a donkey most likely do not seem so different to the average city-dweller. But the offspring of a jack donkey and a mare—a mule—is sterile. (The less common offspring of a stallion and a jenny donkey is called a hinny.) The horse and the donkey, too, are two disparate species.

The geographical isolation scenario seems to explain why there are different species of antelope and so many different species of monkeys. It may or may not explain punctuated equilibrium. If you are an orthodox Darwinist, you will probably say that the changes aren't as abrupt as they seem, and that the geographical isolation theory works just fine. If you adopt a point of view similar to that of Eldredge and Gould, you probably wonder whether it is really all that simple, and find yourself asking whether something else might not have been going on. You may wonder whether the action of natural selection on individuals is all there is to evolution.

Currently accepted evolutionary theory may or may not explain the phenomenon of punctuated equilibrium, but one thing that it cannot explain is the almost-instantaneous appearance of so many different body plans in the Cambrian era. Here we are not dealing with similar species that were altered by isolation. Something else must have been going on. When I say "instantaneous," by the way, I am speaking in terms of geological time, where a "sudden" event may actually have taken tens of millions of years. The earth's geological and biological history is so vast, that a few tens of millions of years does not look like very much when compared to billions.

It may appear that I am concluding this section with more questions than answers. I don't apologize for this. Science, after all, can be defined as an activity by which we ask questions about nature. If we ever discovered all the answers, science would no longer exist. We would be able to concern ourselves only with the history of ideas and with finding new technological applications of some of the things discovered by previous generations.

One reason I think important new discoveries will be produced by the sciences of complexity is that it is a field in which a great number of new questions are being asked. The goal of Tom Ray's Tierra experiments is to find answers to the question, "How did complexity evolve?" Stuart Kauffman developed his theory of auto-catalytic sets in order to find possible answers to the question, "How did life begin?" As research is carried further, the number of questions will multiply. In just a few years, scientists will be asking questions the nature of which we can only guess. Not too long ago, it never would have occurred to anyone that computer simulations might shed light on the problem of the origin of life, or that complex organisms could evolve inside a computer. It is probably impossible to say where the science of complexity will go in the future.

Intelligent Electronic Life

Although Tom Ray's research is fascinating, his speculations are even more so. He has discussed the possible evolution of intelligent life inside computers. According to Ray, if it does evolve, it may be intelligence of a sort that is totally unlike ours, so much so that we will have difficulty understanding it. "Forget the Turing test," Ray says, implying that intelligence in artificial life forms may resemble nothing that we have ever conceived.

The famous "Turing test" was proposed decades ago by the British mathematician Alan Turing. Suppose, Turing said, we wanted to determine whether a super-advanced computer had become truly intelligent. We could perform a test by placing a human in another room and allow him to communicate with the computer. If he could not tell whether he was talking to a computer or to another human, then the computer would have to be judged to be intelligent. Presumably, asking the computer questions about its childhood or about its mother and father would not work, since it would be smart enough to fabricate appropriate answers.

Over the years, the Turing test has been discussed many times in different books and scientific papers. Even those who want to criticize it still pay a certain amount of homage to Turing's idea. But Tom Ray's attitude is a totally irreverent one. He believes that the evolution of artificial life is likely to create phenomena that make the Turing test irrelevant.

Ray believes that electronic life may eventually become intelligent. But we would have trouble recognizing that it exists. Electronic organisms, he points out, do not live in environments which have laws of physics and chemistry like ours. There is no chemistry in a computer environment. Nor do Tierran organisms live in anything that resembles our three-dimensional space. The character of their "space" depends upon the speed with which electronic impulses reach an organism, not on spatial orientations. There is simply no reason, he says, why we should expect the minds of the intelligent digital organisms evolving in Tierra to be anything like ours.

If Ray doesn't think electronic life experiments will necessarily bring about the evolution of alien intelligences with which we will be able to communicate, he nevertheless envisions other benefits that may accrue to scientists during the course of their search for the origins of evolutionary complexity. Electronic organisms, he notes, are only tiny pieces of software. As they become more complex, they may very well turn out to have uses, including some we might never have thought of. He doesn't think that it would be possible to take newly evolved "wild" organisms, stick them into computers, and expect them to do useful work for us, however. They would first have to be "domesticated," he says. We could breed them for our purposes, just as we have bred corn, wheat, dogs, sheep, and cattle. Alternatively, we might decide to engage in a little genetic engineering, and alter wild organisms so that they will perform tasks. Naturally this would all be done with electronic organisms that had not yet attained real intelligence. Making such use of ones that had become self-aware would presumably be something akin to slavery.

If anything like this becomes feasible, it would open up untold possibilities. It is not difficult to think up tasks that we would like computer software to perform. Advanced digital organisms, if they evolve, could likely do much more than that. Consider this analogy. If pigs did not exist, we would not be able to invent the idea of a "pig." It would not be part of our conceptual world. Under such circumstances we naturally would not have bacon, ham, pork sausage, or pigskin. We would never know that we lacked those things. Imagine the difficulty of trying to imagine something like bacon if it didn't exist. Similarly, digitally evolved software could perform tasks the character of which no one has ever conceived. I'm not sure I know what the electronic analogue of bacon is, and most likely no one else does either. But that is exactly the point.

Two Biodiversity Reserves

As an ecologist, Ray spends some time every year in the rainforest of northern Costa Rica. He concerns himself, not only with research involving computers, but also with the preservation of the numerous different species of plants and animals that live in rainforest environments. In recent years, he has become troubled by the destruction of large areas of rainforest in the Sarapiqui region of northern Costa Rica, and has proposed protecting large expanses of the forest. One result of this would be the creation of a biodiversity reserve; preservation of the forest would automatically prevent many plants and animals from becoming extinct.

Ray has already lined up a great deal of support for the project. The Nature Conservancy (TNC) is acting as fiscal agent, and the Costa Rican government has agreed to expropriate land if necessary. The project is supported by a number of conservation organizations, including the Association for Forests and Wildlife, the Organization for Tropical Studies (OTS), the Foundation for the Development of the Central Volcanic Mountain Range (FUNDECOR), the Conservation and Management of Tropical Forests (COMBOS), and the Association for the Environmental Well Being of Sarapiqui (ABAS).

In addition to an organic biodiversity reserve, Ray also proposes setting up a digital reserve. A large and interconnected region of cyberspace would be seeded with digital organisms, which would then evolve through natural selection. If this produces an analog to the Cambrian explosion, very complex organisms might be created. Since electronic organisms can also be viewed as software, a large variety of electronic information-processing organisms would be created. They would, in an important sense, be the analog of the organic biodiversity reserves in the rainforests of the world.

The creation of a digital biodiversity reserve is much more problematical than that of its organic counterpart. To be sure, funding requirements will be smaller. Little will be required beyond the donation of spare CPU time from the computers connected to the project (Ray thinks thousands of them will eventually be linked up). Not that financial support for Ray's work hasn't already been forthcoming. In 1994, the Grateful Dead started things off by performing a benefit concert on behalf of Ray and the digital biodiversity reserve. The Dead's Rex Foundation then awarded Ray a seed grant of $10,000 for the project.

At this point, Ray is nowhere near demonstrating that anything resembling a Cambrian explosion can be made to take place on computer networks. As his projects progress, some interesting results will certainly be obtained. There is no guarantee that high levels of complexity will be created. Even that would not be an unmitigated disaster. Sometimes failures point the way to promising future work. And Tom Ray seems to have the kind of fertile mind that will continue to produce interesting ideas no matter where his work seems to be going.

Sons of Tierra

Naturally there has been widespread interest in Tierra, and in network Tierra, within the scientific community. As a result, numerous different Tierra-like systems have been created by various scientists. As I write this, the most developed of these systems is Avida, which was created by Chris Adami and his colleagues at the W. K. Kellogg Radiation Laboratory at the California Institute of Technology.

Avida and Tierra are similar in many ways, but they are different in one crucial respect. Where the organisms in Tierra live in a world which has a "geometry" that bears little resemblance to the geometry familiar to most of us, those in Avida evolve on a two-dimensional grid. Adami finds that the spatial geometry is conducive to the development of diversity and that it improves adaptive capabilities. He is not attempting to produce a kind of Cambrian explosion as Ray is. His approach is a more mathematical one. He is more interested in testing ideas and predictions about simple living systems, and he analyzes his populations of electronic organisms using the methods of statistical physics. One phenomenon on which he has focused is the evolution of electronic life on different fitness landscapes. He is also interested in obtaining numerical results so that specific ideas about the evolution of life can be tested. Ray's Tierra, you may recall, was not set up with the idea of doing mathematical analyses.

In the network version of Tierra, Ray began with ancestors that were already multicellular. Adami, on the other hand, thinks that experiments with his system may teach us something about the origin of multicellular life. He suspects that, as different species of electronic life appear, individuals with different specializations might form aggregates resembling multicellular organisms.

Adami's work, then, is quite different from Ray's. This suggests that the field of artificial life might soon become a very fertile one. Biological life can be studied from numerous different standpoints. That is why we have such diverse fields as paleontology, ecology, population genetics, molecular biology, botany, and zoology. One shouldn't be surprised if, a few years from now, something similar is happening in the field of artificial life, that different researchers are focusing on different aspects of it, and that they are all achieving significant results.

Adami's Avida software, like that for Tierra, is available to the general public. Adami has written a textbook on artificial life. It comes with a CD-ROM containing versions of Avida for a variety of different computer platforms. See the bibliography at the end of this book for more details. The software can also be downloaded from Adami's website. See the "World Wide Web Resources" section at the end of this book.

Varieties of Artificial Life

Studies in artificial life have already transcended the bounds of "electronic biology." I'll be citing a variety of different examples later. For now, I'll confine myself to mentioning one that also deals with evolution, though not of the biological kind. Anthropologist Nicholas Gessler of the University of California at Los Angeles believes that cultural evolution, too, can be studied on a computer. He comments that "investigations may help to clarify the role of the individual in society, the dynamics of cooperation, of deception and manipulation, of the usage of language, and of the overall organization of culture."

As an anthropologist, Gessler has spent over fifteen years working with native communities on the Northwest Coast. He has researched records of cultural change in the eighteenth and nineteenth centuries, when Native Americans came into contact with European cultures (among others; there also seem to have been influences from the Sandwich Islands and from Canton). Around the mid-1990s, he began to wonder whether work in the sciences of complexity might not be able to shed additional light on what was going on.

I won't go into Gessler's ideas in any great detail. His work lies somewhat beyond the scope of this book, which deals with the ways in which complexity research can shed light on the nature of life. However, the suggestions that Gessler makes cause one to suspect that a whole world of new possibilities may lie ahead. It can be argued that human societies—real or modeled in a computer—are a form of artificial life. If biological and cultural evolution can be modeled on computers, why not go on to technological evolution (will computer organisms begin to invent new technology before we do?), to economics and to other fields as well? Unlike studies in evolutionary biology, research in "electronic biology" seems to have implications that are far-ranging indeed.

8

SWARM

When Tom Ray devised Tierra in 1989, this was not the first time that artificial life had been created in a computer. Work in the field of artificial life had been going on for many years, but most often it was done by scientists who worked alone, unaware of what other people in the newly emerging discipline were doing. They were isolated from one another because there was no journal devoted to artificial life, and no conferences on the subject. When Santa Fe Institute computer scientist and artificial life pioneer Chris Langton organized the first artificial life conference at Santa Fe in 1987, he had difficulty discovering who the leading people in the field were. The problem was, they published their scientific papers in any journal that would accept their work. Artificial life papers, Langton joked, could appear almost anywhere, even in the *Italian Journal of Basket Weaving*.

In 1987, no one had created artificial organisms that could reproduce, mutate, and evolve. However, one should not conclude that the work being done then was unimportant. Artificial life scientists concentrated on making computer organisms that mimicked certain aspects of the behavior of living creatures. They also created models that were intended to illuminate the workings of Darwinian evolution. Artificial life was already a lively field of endeavor. Only a few people were aware of this fact, though.

One of the more striking artificial life programs that already existed in 1987 was computer scientist Craig Reynolds' "Boids" (short for "birdoids"). Reynolds' boids engaged in flocking behavior that was uncannily similar to the flocking of real birds. His artificial flying creatures followed simple rules that governed the behavior of each boid as an individual. There was nothing in the program that molded the behavior of the flock as a whole. Yet the flocks were able to engage in some very sophisticated behavior. If the flock encountered an obstacle, for example, it would split into two groups. Some of the boids would fly to the left of the obstacle, some to the right. The flock would then reform on the other side.

Though the boids existed only in a computer, Reynolds' program was not only fun to watch (consult the "World Wide Web Resources" section to learn where to look at it online or download a copy), it seemed to create insights into the flocking behavior of real animals. There was no guarantee that real birds used the same rules as the ones employed in Reynolds' program. But the behavior of the boids looked so real that ornithologists began asking what the rules were. The program caught the interest of animators, who used boids as the basis for software that showed flocking bats in "Batman Returns" and a wildebeest stampede in "The Lion King."

Flocking boids, incidentally, are a good example of the phenomenon of emergence. The rules which Reynolds devised to control the behavior of his artificial organisms were very simple ones that applied only to the flight patterns of individual boids. The rules were:

1. Separation: steer to avoid crowding local flock mates.
2. Alignment: steer toward the average heading of local flock mates.
3. Cohesion: steer to move toward the average position of local flock mates.

Note that none of this specifies anything about collective behavior. But flocking appears nevertheless. Once again, we find that complex behavior has emerged out of the interaction of simple elements.

The scientists who do research in the field of artificial life are in effect creating artificial worlds. Sometimes these worlds are very simple; some of them have only one spatial dimension. Others are more complex. An example would be the world being created by Tom Ray in the network Tierra project. Yet another would be SimLife, a game created by Maxis, the developers of Sim City. One can populate a SimLife world with different kinds of plants and animals, specify their mutation rates, tinker with their genetic code, and watch them evolve. It is this complexity, incidentally, that makes SimLife a game (one that is, incidentally, somewhat frustrating to play) rather than a tool that can be used for serious scientific research. As I pointed out previously, in scientific experiments it is necessary to hone in on the essential features of a phenomenon. If matters become too complicated, then it is difficult or impossible to determine what is going on. In SimLife, complexity does not emerge from the application of simple rules. The elements of the game—the environment and the artificial creatures that inhabit it—are complex to begin with.

Little is required to create an artificial world. It is only necessary to devise electronic life of one variety or another and to place it in some suitable computer environment. If everything has been set up properly, then it is possible to watch this world and/or its creatures evolve. An artificial world does not have to embody Darwinian evolution. For example, the scientists who model ecosystems generally do not include the possibility of evolution at all. They are more interested in interactions between the artificial creatures that inhabit their worlds. Similarly, the scientists who create models of human societies do not take evolution into account. They are more likely to be concerned with such problems as the relation between agricultural surpluses, or the lack thereof, and the political structure of a preliterate society.

Artificial worlds are encountered in many of the computer games marketed nowadays. The inhabitants of these worlds are often capable of diverse kinds of behavior. If the game involves combat, for example, the player's opponents may decide to either stand their ground, charge, or run. Such games have no scientific value. The creatures that inhabit these worlds do not interact with one another, or modify their behavior. On the contrary, it is the player who must learn how best to interact with them.

Sometimes, of course, worlds come to an end. An experiment run on Echo, an ecological simulation program developed by University of Michigan computer scientist John Holland in the mid-1990s, provides a good example. In one simulation, the world of Echo was populated by three different kinds of creatures: caterpillars, ants, and flies. The parasitical flies preyed upon the caterpillars. But the latter had a defense. They secreted a sweet substance that the ants liked to feed upon. Since the flies were allergic to the ants, caterpillar mortality was not excessively great. Then one of the flies mutated to become a more efficient predator. Its genes spread throughout the fly population (Echo was an ecological simulation program that did include evolution), and the number of caterpillars began to decline precipitously. This was followed by a fall in the number of ants, which were soon on the edge of extinction. The result was ecological catastrophe.

This is not the only kind of catastrophe that can occur. Species can also become extinct in ecological simulations which do not have evolution. If the predators are a little too efficient, for example, they are likely to decimate the ranks of their prey. Something like this occurred in the Echo simulation described above, although it was evolution that was the ultimate cause of the ecological imbalance. Similarly, two or more different species of organism may compete for the same resources. If one of them is too good at what it does, the others may find it impossible to compete. Such things only make artificial life models more realistic. After all, the same things can happen in "real life."

"SSWOOORMM"

One of the scientists doing research in the artificial life field in 1987, the year of the first conference on the subject, was, naturally, Chris Langton. He was one of the pioneers in the field. Although his scientific work was not exactly trivial, Langton was soon to become best known as an organizer. In fact he has been called the "father of artificial life" at times. (For more on Langston's work in artificial life, see the Steven Levy reference in the Bibliography.) Although he was not the first to do research in the field, there is some justification for this characterization. It was Langton who helped artificial life researchers gain contact with one another, and who was instrumental in fashioning the field into a respectable scientific specialty. The latter was an important achievement. When one scientist mentioned

Langton's first artificial life conference to a noted scientific friend and asked whether he should attend, he was advised that he might as well go, but that it was wisest to tell no one that he had been there. It is by no means surprising that an eminent scientist would have expressed such an opinion in 1987, but it is very unlikely anyone would give such advice today.

When French journalist Pierre–Yves Frei visited the Santa Fe Institute in 1995, he was especially intrigued by Swarm, a computer software system that Langton and his colleagues were developing. His interest even extended to the pronunciation of the name. "Swarm," he told his readers in an article published the following year, was pronounced with *"une bonne grosse dose d'accent américain"* (a nice big dose of American accent). A good rendition of the name, he went on, might be "SSWOOORMM." But this was only the first of Frei's superlatives. Swarm, he went on, was "alive, massively alive!"

Of course Swarm wasn't really alive. It is not an artificial life simulation, but rather a software system that is designed to make it easier to create artificial worlds. It is something like the standardized equipment used by experimental scientists. At one time, scientists who wanted to perform an experiment had to blow their own glass, wire their own circuits, and so on. Today, more often than not, they can buy standardized equipment from companies that specialize in it, and concentrate on the business of doing experiments. Swarm is something similar. The scientist who wants to model living systems in a computer no longer has to write all his own programs. He can download Swarm from the Internet and use it to create almost any kind of artificial world that he desires.

To better understand Langton's motivations for creating Swarm and learn what might be accomplished with it, it is necessary to know a little about how research was conducted in the artificial life field before 1995. The majority of the scientists who did research in this new field were not computer scientists. They were people who had picked up bits of programming knowledge here and there as they needed it. Their programs were not as well-designed as they could have been. It was not always possible to determine whether certain results were really meaningful, or whether they were caused by some quirk in the software.

There were other problems too. If an artificial life scientist wanted to carry out a second experiment, he would generally have to design new software to create a new artificial world. It was hard to compare one's work with that done by other scientists. For example, suppose that I create a world using my software, and you create a similar world using yours. You obtain certain results. I obtain results which are similar, but not quite the same. Was the disparity caused by the differences in the software we were using? Or is it an indication of something more fundamental that should be investigated further? There is no way to know.

Scientific results are generally not accepted until the work done by one scientist—or group of scientists—is duplicated in the laboratory of another. Suppose you claim to have discovered a new drug that is extraordinarily effective against the AIDS virus. It is perfectly reasonable for me to remain skeptical until your work is duplicated by someone else. Perhaps I say that I have measured the energy levels of the nucleus of the carbon atom to an unprecedented degree of accuracy. Again, you may continue to entertain reservations until other scientists have confirmed my results. Science, as we all have heard many times, is supposed to be objective. If one scientist observes a certain phenomenon, then others ought to be able to see it too if they perform the appropriate experiments in their own laboratories. It was the lack of confirmation that put an end to work in cold fusion. Scientists who tried to duplicate the original experiments in their own laboratories were simply unable to obtain any significant results.

Perhaps this issue is less important in the field of artificial life than in most others. Artificial life scientists, after all, do not observe the real world. They create computer models in order to better understand certain of its characteristics. At best, their results can only be suggestive about the processes that go on in environments populated by living biological creatures. Nevertheless, there are times when it would be useful to compare the results obtained by one researcher with those of another. The importance of this cannot be stressed too much. To give yet another example: If you say that punctuated equilibrium is observed when the creatures living in system "A" attain a certain degree of complexity, I will not be able to check your results if I use software system "B."

Swarms

Suppose software could be developed that could be used by scientists carrying out many different kinds of artificial life experiments? If it could, they would be relieved of the responsibility of having to laboriously piece together their own software. They would not have to write new programs every time they wanted to carry out a new experiment. Comparing results would be a much less difficult task.

Why not attempt to create something that could be used for other purposes than studies of artificial organisms? Such a system might be used by economists, for example. A stock or commodities market, after all, is composed of individuals who buy and sell. They thereby create a market, which seems almost to take on a life of its own. This is reflected in the commentary of financial journalists, who often attribute human-like emotions to the stock market. We've all heard this kind of talk. It is sometimes said that the market is "nervous" or "elated" or that it has recently exhibited some other characteristic. They may not be exaggerating as much as one might think. Some researchers think that markets—and computer simulations of them—are indeed a form of artificial life.

There are many other kinds of simulations that could be usefully simulated in Swarm. One could use it to model mammalian immune systems on a computer, for example. An immune system could be represented by an artificial world populated by cells, antibodies, and some sort of invading organism. Swarm could also be used for analyses of the evolution of human societies and of such mundane problems as traffic congestion. Traffic flow, after all, is nothing more than the behavior of a swarm of automobiles.

The name "Swarm" is a very descriptive one. A Swarm world is made up of a swarm of "agents" and the environment that they populate. An ant simulation, for example, would consist of a swarm of electronic "ants," their environment, and a set of rules to govern their behavior. A swarm of artificial bees could be constructed in a similar manner. Alternatively, one might construct a simulation in which more than one kind of agent exists. An ecological simulation, for example, might be made up of plants, the herbivores which eat them, and the carnivores which in turn prey upon the herbivores.

It is also possible to create swarms within swarms. For example, a pond might contain a swarm of one-celled animals. These animals, in turn, could be regarded as being made up of swarms: the organelles (nucleus, mitochondria, and so on) within the cells. Furthermore, the organelles themselves could be viewed as swarms; they are made up of proteins, DNA, and other organic molecules. In all these cases, the elements of a swarm are called "agents." Something that is a swarm at one level may also be an agent at a higher level of complexity.

I don't think that anyone really has any idea where the concept of swarms within swarms might lead. At this writing, numerous different experimental groups have expressed interest in the Swarm simulation system. They have only begun to carry out their research, and little has been published. Although there have been some postings on the World Wide Web, I have not run across any references to papers published in scientific journals. There is nothing surprising about this. Work on Swarm began in 1995, but it took a couple of years to develop it into a fully featured system that could make a variety of different artificial worlds available. New scientific findings are not produced overnight, no matter how efficient the system on which they are run.

Improvements continue to be made. There now exists a Swarm "library" from which each individual researcher can select a world which suits his requirements. These range from two-dimensional worlds in which agents can move about to models that represent telecommunication networks on which agents can send and receive messages and trade stocks or commodities. Swarm also provides a variety of tools which can be used to design a world and analyze the behavior of the swarm within it. For example, if the agents in a Swarm world evolve, one would want to be able to analyze their genetic makeup and determine how they differ from their predecessors. In an ecological simulation, it is necessary to be able to tell how many electronic creatures of each variety exist at any given time. With Swarm, tools for looking at such things no longer have to be created by the individual scientist.

I have been talking about swarms rather than the agents that comprise them because Swarm is a tool for studying collective behavior when agents interact with one another and with their environment. It is at the level of the swarm, not that of the individual, that matters begin to become interesting. Matters could hardly be otherwise. After all, the scientists who use Swarm are interested in observing the collective, emergent behavior that appears when a swarm is created.

We have already seen one example of swarming in Craig Reynolds' Boids (which does not make use of the Swarm platform). Examples abound in the real biological world. An ant or bee colony, or a simulation of an ant or bee colony for that matter, is capable of engaging in complex behavior. Yet individual ants have simple nervous systems, and their repertoire of behavior is extremely limited. One even sees swarm behavior in human beings, who are infinitely more complex. It has been long recognized that the psychology of a crowd is not the same as the psychology of individual human beings. One might think of lynch mobs here. However, collective behavior appears in many different contexts, most of them less violent. Attending a sports event can generate an excitement that is absent when one watches a game on television. Investors in the stock market act collectively, and produce behavior that one does not see in them individually. This can, for example, cause a market to crash when no one wants it to crash. All this suggests that making computer simulations of different kinds of swarms could produce results that are very revealing. There are swarms on all levels. If swarms were not more than the sums of their parts, we would not, for example, have human societies.

According to Langton, "what makes swarms scientifically interesting . . . is the coupling between the individual and group behaviors." Although the individuals are usually relatively simple, their collective behavior can be quite complex. Swarms allow us to focus directly on the roots of complexity: the simulations capture simplicity as it is in the process of generating complexity.

Like Tierra and Avida, Swarm is available free to the general public. It was originally written for the Unix operating system, but it can be run on a PC or a Macintosh. To do this, it is necessary to obtain Linux, an operating system which mimics Unix, and which is designed for personal computers*. You will need Linux to run Swarm on a Macintosh, and to run artificial life programs that are not available in Windows or Macintosh formats. Linux was developed by the Finnish computer scientist Linus Torvalds when he was a 21-year-old student. He couldn't afford the thousands of dollars necessary to obtain Unix for his PC. He solved the problem by writing software of his own. He posted it on the Internet, asking for suggestions for improvements, and gave away the final product. Linux can be downloaded free from a number of different sites on the World Wide Web. Once it is installed it is possible to switch back and forth between Linux and operating systems such as Windows. I don't especially recommend that you get it, however, unless you're much more dedicated than I am. I have been told that if you know what you're doing, you can probably install Linux in a day. If you don't, it will take much longer. Once you're done, it is still necessary to install Swarm. Scientists get paid for doing this kind of thing. Most of the rest of us don't.

Research Using Swarm

As I write this, Swarm is still being developed. A number of different versions have been released since the prototype was first made available, and the system has been given additional capabilities. Thus it should come as no surprise that a great number of research projects which make use of Swarm are underway. They may not be written up in the journals yet. However, it is possible to obtain all the information that one might want about the projects in postings on the World Wide Web. While I make frequent references to different kinds of software and to the World Wide Web, I am not a computer expert. In the last few years, the Web has become a major vehicle for the dissemination of scientific information.

---

*Since I wrote this, a Windows 95 version of Swarm has become available.

At the moment, there are a great number of works in progress. I have encountered nothing that resembles a project like Tom Ray's Tierra. As far as I can tell, Swarm is not yet being much used to study evolution. Few of the initial research projects are concerned with biology at all. Those that are biological in nature tend to be studies in ecology.

One should not conclude that the work with Swarm is unrelated to artificial life. In a sense, everything done with Swarm *is* Alife. When economists use Swarm to model financial markets, the agents they create represent the individual human beings who do the buying and selling. Such simulations really do not differ from ecological simulations in any fundamental ways. In both situations, the big animals (or players in the market) sometimes eat the small ones.

Interestingly, economics is one of the areas in which artificial life has achieved its greatest success. Traditional economics, which makes use of such abstract concepts as "supply," "demand," and "markets," has naturally had its successes too. Without modern economics, the Federal Reserve Board would not know how to regulate the economy, and we would still have the "boom and bust" cycles that were so characteristic of nineteenth-century capitalism. Without modern economics, the International Monetary Fund would not know what conditions it should attach when it made loans to governments experiencing financial difficulties.

However, traditional economics makes use of abstractions which have no counterparts in reality. It conceives of perfect markets, perfect information, and perfectly rational agents. In this respect, it has not progressed far beyond Adam Smith, who, in his 1776 book *Wealth of Nations*, spoke of an "invisible hand" which brought supply and demand into equilibrium with one another. Traditional economics simply has no way of dealing with the actions of individuals, or the effects of these actions on markets. It still makes use of Smith-like abstractions. These allow one to create workable models of economic reality much of the time. There are situations in which the endeavor is likely to fail. Modern economists realize this. Hence the emphasis on "micro-economics" in recent years.

Using artificial life simulations allows scientists to model economic behavior at a deeper level than traditional economics can. There has been a great deal of interest in artificial models. The

Swarm-based research projects currently underway include, not just the creation of models of markets, but also models of manufacturing systems, models of the exploitation of such renewable resources as fisheries, and a study of pollution. The last, naturally, has economic as well as ecological aspects. An industry will not pollute if economic incentives and disincentives make it unprofitable to do so. And instituting pollution controls has an effect on profits. Among other things, this has affected the price of books. Paper prices rose dramatically when pollution controls were instituted in the paper industry a number of years ago.

A number of Swarm projects deal with more traditional kinds of artificial life. For example, David Sumter of the University of Manchester Institute of Science and Technology is using Swarm to study the collective behavior of honeybees, and Claudia Pahl–Wostl's group at the Swiss Federal Institute of Environmental Science and Technology has been using a Swarm-based model to generate ecological networks. Meanwhile, Matt Hare of the Macaulay Land Use Research Institute is creating a model of red grouse population dynamics.

Swarm can be used to improve existing simulations as well as to create new ones. Alex Lancaster, a computer scientist at the Santa Fe Institute, has created a model called "Bugverse" to study neural networks embedded in creatures which inhabit an artificial ecology. As they adapt to their environment, connections between their neurons change, altering their behavior. The bugs' networks are built in a modular manner; a group of neurons at one level becomes the basic building block of a complex at the next higher level. The idea is to test hypotheses about the evolution of brain complexity. Bugverse is the successor of a model called "Bugworld." Using Swarm for the former allows more complexity to be built into the model.

Swarm and Archaeology

It is easy to see that a system such as Swarm should be well adapted to modeling the behavior of such creatures as ants and bees. After all, they are simple organisms which interact with one another in a simple manner. It can be somewhat surprising to see that it has been adapted so successfully as a platform for models of human behavior. After all, we aren't simple "bugs" who interact according to a small number of basic rules. Our behavior is influenced by our beliefs, fears, desires, and by our life experience as well.

Nevertheless there seem to be some aspects of human behavior which can be successfully handled. For example, an agent participating in a market can do only two things: buy or sell. The agents must act according to different "expectations," or they would all be buying or selling at the same time. It is easy to devise simple rules according to which agents "predict" market activity differently. One can then see not only how this activity causes price fluctuations, but also determine which strategies are most effective.

Similarly, one can apply Alife techniques to problems in archaeology if some set of simple rules can be devised. Archaeologist Tim Kohler of Washington State University and Santa Fe Institute computer scientist Eric Carr are developing such a model to use in testing various anthropological theories about Anasazi village formation in the Verde District during the period from A.D. 901 to A.D. 1300.

The Anasazis (from the Navajo, meaning "ancient ones") were a native American people who developed a complex culture that lasted from around A.D. 100 to modern times. They lived in parts of what are now Utah and Colorado, and in most of New Mexico and Arizona. During the most vigorous phase of the culture, which lasted from around A.D. 700 to around A.D. 1200, the Anasazis built large, permanent villages which included multiroom and multilevel buildings. The cultural artifacts that have been unearthed indicate that their artistic craftsmanship reached a high level, and their religious ceremonies became quite complex.

One might think it would be impossible, or at least very difficult, to model such a culture on a computer. Indeed it would be. Kohler's and Carr's goals are more modest. Using Swarm, they are engaged upon a long-term project that is intended to relate Anasazi village formation to food supplies. The Anasazis were an agricultural people who depended heavily upon the cultivation of maize and beans. The size of the agricultural areas that they occupied varied with fluctuations in climate that affected their ability to cultivate marginal agricultural land.

But Kohler's and Carr's model is not just about climate. In their simulation, the basic units—the agents—are Anasazi households. The households are assumed to operate according to simple rules involving the sharing or hoarding of food. The model relates this behavior to food supplies and climate fluctuations. Kohler and Carr hope that a reasonable picture of village formation will emerge. In other words, they hope to see complexity arise from simple interactions.

Since this is a long-term project, it is not yet apparent how well their goals will be realized. The creation of such a model supports the idea that artificial life techniques might successfully be used to study interactions within human societies.

But What Does All This Have to Do with Life?

If Swarm can be used to analyze prehistoric societies, financial markets, and such things as recreational boat traffic (one such study is currently underway), this is all very well. But what does this have to do with the problems that have been discussed in this book, such as the origin and evolution of life and the evolution of biological complexity?

The above examples illustrate Swarm's flexibility. Any kind of artificial life simulation can be set up on the system. It is reasonable to assume that many different kinds will be. Scientists who create computer models are relieved of much of the onerous task of programming and are free to concentrate upon their own scientific goals. This has allowed some to venture into previously unexplored areas, such as village formation. It will certainly allow others to create meaningful models for the origin and evolution of life.

I think it is apparent that many of the initial modeling projects do not utilize Swarm's potential to the fullest. This is only to be expected. When a new scientific tool is developed, scientists generally use it to study the simpler problems first, and then the more complex ones. If the initial experiments were too complicated, it would be hard to understand what was going on in them. Once these experiments are concluded, then a greater number of factors can be included. This is the surest path toward scientific progress.

Some of the applications of Swarm are even more simple than their predecessors. Gecko, an ecological modeling system based on the Echo system, does not include the possibility of evolution. This is reasonable enough, because Echo and Gecko are used to study ecosystem dynamics, in which evolution is of little or no importance. The coevolution of predators and prey is an interesting subject, but there is nothing unreasonable about trying to gain an understanding of the nonevolutionary aspects of predator–prey interactions first.

If I were to try and predict what kinds of experiments Swarm might be used for in the future, I would be on somewhat shaky ground. However, there is nothing wrong with mentioning a few possibilities. It should be possible to use Swarm to create a detailed model of a Kauffman autocatalytic set, for example. It would be necessary to create some kind of artificial chemistry from which artificial life might spring. But artificial chemistry is a topic that has already been explored, most recently by Barry McMullin, Director of the Artificial Life Laboratory at Dublin City University. McMullin became interested in artificial chemistry after learning about some theories espoused by the Chilean biologist Francisco Varela. The techniques that he and other scientists are developing could presumably be applied to autocatalytic sets and Kauffman landscapes (no one calls them "Kauffman's fitness landscapes" anymore), among other things.

I have not encountered reports of any research projects which exploit Swarm's ability to create swarms within swarms. I cannot help but wonder whether this technique might not, in the future, be successfully applied to the study of the evolution of multicellular life. The first multicelled organisms presumably evolved from swarms of single-celled creatures. An agglomeration of single-celled animals evolving toward multicellularity would itself be part of a swarm. The hierarchy doesn't stop there. A species is a swarm of individuals. If species really do play a unique role in evolution, studying models of them in Swarm might produce some fruitful ideas.

The possibilities seem endless. There are undoubtedly many that no one has yet thought of. One cannot help but have the impression that we are witnessing the beginning of a series of new explorations of the possibilities inherent in artificial life.

The Fate of Creative Scientists

During the greater part of this chapter, I concentrated on descriptions of individual Swarm experiments. I have not said very much about the career of Chris Langton, who initiated the project, and who is now the director of the project.

Langton is largely responsible for the creation of Swarm and the one who probably did the most to gain recognition for artificial intelligence as a legitimate scientific field when he first organized the artificial life conferences, but he does not think of himself as an administrator, rather as a scientist, or perhaps not even a scientist. "I consider biology to be part of physics and cosmology," he said to me recently. "I'm more of a natural philosopher than a specialist in any particular discipline, and I tend to follow my scientific nose across many disciplinary boundaries."

Langton began his research in the field of artificial life during the early 1980s. By the time that he began work on Swarm, he was considered to be an important figure in the field. He has temporarily put this work aside in order to direct the Swarm project. I do not think that he is particularly happy in this role. It is a common fate of creative scientists to become administrators during their later years. Langton, however, is still relatively young, perhaps too young to sink permanently into an administrative job. But it doesn't necessarily follow that he will get back to devoting himself entirely to scientific work again. Among other things, he is currently starting a company to apply Swarm to large-scale complex problems in industry.

Well, that's an aspect of artificial life too.

9

THE PROMISE OF COMPLEXITY

A ttempting to follow the work being done in the sciences of complexity can be simultaneously exciting and frustrating. A great deal of scientific work has been done, but it always seems to be just on the verge of answering the question "What is Life?" The research already performed has not given any definitive answers to the question, "How did life begin?" Stuart Kauffman's theory of autocatalytic sets is certainly brilliant and compelling. But Kauffman's idea has little or no experimental confirmation. Similarly, his work on gene interactions and their relationship to fitness landscapes might answer some important questions in molecular biology. Again, it is hard to relate Kauffman's research to work that is being carried out in the laboratory.

The work being done by Julius Rebek and by Reza Ghadiri is quite innovative, and might eventually provide us with some answers. In particular, Ghadiri's research might shed some light on the ways in which complexity in biological and prebiological systems initially evolved. However successful his experiments may be, they will obviously not allow us to travel back in time to see what actually did happen. Even though their work is experimental, there are aspects of it that would be hard to confirm by experiment!

Many questions will be answered if, at some point, we are able to study extraterrestrial life. If we were able to see how other life-forms were constructed, it might be possible to deduce what properties a living system must have, and which properties are accidental. Since we have not yet found life elsewhere (the claim that there was once life on Mars is very controversial), we are limited to observing life inside computers.

It is fair to say that research in the sciences of complexity has not answered any of the enduring questions about life. We don't know how life began. We don't know whether or not natural selection is the only important factor in evolution. Are there laws of complexity that constrain the ways in which life can evolve? No one is really sure. There are questions about the significance of punctuated equilibrium. Should the Cambrian explosion be replayed, we don't know whether the organisms that evolved would be pretty much the same or something entirely different. For that matter, how does complexity in living beings evolve in the first place? What is it exactly that distinguishes something that is "alive" from something that is not? Most of the time we have no difficulty telling the difference. We would all agree that a squirrel is a living creature, while a rock is not. What about such borderline cases as viruses? Should Julius Rebek's synthetic chemicals be regarded as being alive in some sense? What about the electronic organisms that live inside computers? Are they alive? Even researchers in artificial life do not agree on this last question. Some would say that these organisms are indeed living creatures, while others regard them as nothing but tiny computer programs.

Scientific Success

I don't mean to paint a pessimistic picture of the progress that has been made in the sciences of complexity. Traditional biology cannot answer the questions that I have cited. When viewed from a purely scientific point of view, the new knowledge gained by complexity scientists has been enormous, as we have seen.

Somewhat paradoxically, the greatest practical successes have come in such areas as economics and in the prediction of the behavior of financial markets. In 1991, Doyne Farmer, a physicist associated with the Santa Fe Institute, and Norman Packard, another physicist who had known Farmer since their student days, founded the

Prediction Company with the intention of developing forecasting methods for market prediction and the trading of financial instruments. Using techniques developed in chaos theory and in the sciences of complexity, they found that they could create mathematical models which would indeed allow a trader to make money. They realized that their computer models did not have to be right all the time. The ability to make accurate predictions 55 percent of the time is quite enough to create profits.

Farmer and Packard found that, although financial markets may seem to be chaotic most of the time and long-term predictions may be difficult or impossible, they could find significant short-term patterns. In other words, they couldn't say what a market was going to be doing next month or even next week, but they could see little "pockets of predictability" appear here and there.

Farmer's and Packard's method has apparently been quite successful. They now have the backing of the Swiss Bank Corporation, a major international financial institution. They are not allowed, however, to divulge how much they trade, or where they do it. Their contract with the corporation prohibits them from doing so.

Farmer's and Packard's company is operating successfully and making money. According to Farmer, the studies that he and Packard have done prove that, "by rigorous scientific standards," markets can be beaten. "We have found statistically significant patterns in financial data," he says, and adds that he would like to write a technical book on the subject. Given the amounts of money involved, he suspects that his and Packard's partners will never allow them to do this. He does believe that the work done at his company has led to the development of "prediction machinery" that would allow prediction of a lot of things, such as the weather, global climate patterns, and the spread of epidemics.

Farmer remains a scientist in spite of his and Packard's success. He says he would like to score "big time" in the financial markets and then move on to more interesting problems, such as artificial life, artificial evolution, and artificial intelligence. Farmer and Packard do not see making money as an end in itself. If they did, they probably would not have become scientists in the first place. To be sure, they are currently engaged in making "bets" with large sums of money. They regard the development of the basic principles of prediction to be more important.

Controversy

The study of complexity has apparently led to the development of ways to beat the market. Does it have achievements that are clearcut and significant? This seems to be a matter on which there is not any general agreement. In an article published in *Technology Review*, Robert J. Crawford, director of the Office for Sponsored Research at Harvard University, wrote that "alife is an artificial concept whose usefulness, at least at this stage, is largely in the eye of the beholder."

Crawford was writing specifically about Alife. His comments could easily be extended to studies in the science of complexity in general. Alife is, after all, just a subdiscipline of the sciences of complexity. However, this book is about life, both the biological and electronic varieties, so I will only quote comments on research in artificial life in the paragraphs that follow.

It is interesting to note that, although the field of Alife has attracted mathematicians, physicists, and computer scientists, professional biologists have for the most part remained aloof. It is true that Tom Ray is an ecologist. He is one of the exceptions. Biologists frequently show a lack of interest in the results achieved by practitioners of Alife. At times, they have expressed disdain, and have sometimes expressed the opinion that Alife is only an electronic game not relevant to the real world.

Harvard paleontologist Tomasz Baumiller once asked, "Where's the biological reality behind what [the artificial life researchers] do?" He says that he does not see how creating organisms that are made of nothing but computer code can cast new light on the fossil record he studies. His colleague, Harvard biologist Richard Lewontin, has expressed a similar point of view. He says that Alife "hasn't taught me anything yet," but admits that he wouldn't deny that it might in the future. He does say, however, that Alife has not come up with any of the kinds of surprises that one encounters in real-world biology, such as the recent discovery of a megabacterium that is large enough to be visible to the naked eye.

Perhaps some skepticism among biologists is to be expected. For one thing, they do not ordinarily work a lot with computers, as scientists in certain other fields do. When they do make use of computer technology it is because computers make it easier to store and organize data. Physicists, on the other hand, tend to be more sympathetic. In Einstein's day, theoretical physicists worked with pencil and paper. Today, they are likely to use computers to solve equations and do theoretical analysis. They also seem to take naturally to ideas in the field of complexity because much of what is done has similarities to their own work. Some work in physics—condensed matter physics, for example—*is* research in the sciences of complexity. Condensed matter physics deals with substances in the liquid or solid states. This is an area in which complexity abounds.

Physicists, if they have commented on Alife research at all, have tended to be enthusiastic about it. However, some criticisms of Alife research has come from scientists who work in the sciences of complexity themselves. Brandeis University computer scientist Maja Mataric, who studies cooperative movement in robotic insects, believes that Alife computer models "vastly oversimplify the real world." Producing truly complex behavior, she says, requires "noise" and perturbations which are not yet part of Alife simulations.

If there is one thread that runs through all these comments, it is the idea that you only get out what you put in. The critics say that artificial life computer models may tell us something about the workings of complex computer programs, but not much more. They couldn't; there is too much in models that is arbitrary, and there is no way of knowing precisely what features of life one should attempt to capture in a computer program. Whatever Alife is about, in other words, certainly isn't life. There is an enormous difference, they say, between crunching numbers and studying fossils or living organisms.

Naturally, not all biologists express such points of view. I think it is clear there are large differences between the outlooks of most complexity scientists and the outlooks of most of the people working in the biological sciences. Stuart Kauffman has made this point by commenting that biologists often think the term "theoretical biology" is an oxymoron. The controversy is likely to continue for some time. On the one hand, there are scientists who believe that Alife has the potential to tell us a great deal about the origin and evolution of life. On the other, there are equally competent scientists who think that it has so far proved to be largely irrelevant.

I don't think that the persistence of such controversies will prove to be inimical to the development of the field of artificial life. Scientists often develop opposing points of view, and the resulting debates tend to enrich science. Sometimes it is possible to gain perspective by looking at what happened in another field of scientific endeavor. This makes it possible to look at the "broad picture," and to forget the details of a controversy in which one may feel personally involved.

During the late 1920s, Albert Einstein and the Danish physicist Niels Bohr began to debate with one another about the nature of quantum mechanics. It was an argument that was to last nearly a lifetime. Einstein was disturbed by the indeterminacy and probabilistic interpretations that seemed to have become a part of quantum mechanical theory. He objected to the point of view developed by Bohr and his colleagues at the Institute for Theoretical Physics in Copenhagen and never passed up a chance to argue about it.

According to the "Copenhagen interpretation" of quantum mechanics, probability and uncertainty are fundamental characteristics of the subatomic world. One cannot predict when a radioactive atom will decay, and one cannot predict when, or in what direction, an atom that possesses the requisite energy will emit a photon. One can describe the position of an electron in terms of probabilities, but one cannot precisely define its position in space.

Einstein did not dispute that quantum mechanics gave results that could be confirmed by experiment to a very high degree of accuracy. Both he and Bohr realized that, most of the time, one dealt not with one particle or with one atom, but with billions of billions of billions of them. This caused the probabilities to average out. What Einstein objected to was Bohr's idea that, at its most fundamental level, the world was governed by indeterminacy. He believed physicists would eventually find a theory to describe the behavior of matter in a completely deterministic way. "God does not throw dice," he would say again and again as he and Bohr argued. At one point, Bohr became quite exasperated and admonished Einstein to stop telling God what to do.

Einstein lost the argument in the sense that the Bohr interpretation is the most commonly accepted view of quantum mechanics today. But he never gave up his opposition, and this forced physicists to think about matters that they might otherwise never have considered. This enhanced the development of modern physics. It still does. It is likely, for example, that some of the ideas developed by the British physicist Stephen Hawking would never have been put forward if Einstein had stopped arguing and had gone on to other matters.

Similarly, if the critics of artificial life were not constantly talking about its irrelevancy, it could very easy become irrelevant. Having unbelievers look over your shoulder is sometimes just what is needed to spur you to produce more rigorous work. A sense of enthusiasm is often not enough. The enthusiasm can too easily become exaggerated. It seems to me that work in the field of artificial intelligence would have progressed more rapidly if the first researchers had not been prone, at first, to making extravagant claims. They spoke of real intelligence within computers—and then discovered they were unable to program machines to translate languages or to play a decent game of chess. Today there exist both chess-playing and translation programs that carry out their assigned tasks reasonably well. One can buy computer chess software capable of beating most masters for under $50. As such programs have been developed, scientists have begun to understand how wide the gap is between real intelligence and being able to program certain specific abilities into a computer. Deep Blue may have beaten world champion Garry Kasparov at chess, but it wouldn't know how to converse with a human being while ordering a meal in a restaurant.

So I don't think it will do any harm if a damper is occasionally put on the claims of the advocates of complexity. This will only induce them to look for ways of justifying these claims. That could lead to the creation of some very interesting science.

Artificial Life as Philosophy

Daniel Dennett has expressed optimism as about the future of artificial life. He thinks that it could become a kind of philosophy in which thought experiments could be created and tested in a rigorous way. In an essay published in the first issue of Chris Langton's journal *Artificial Life*, he opined that artificial life techniques could help philosophers devise thought experiments that would be "kept honest by requirements that could never be imposed on the naked mind of a human thinker acting alone."

Thought experiments have been used quite often in philosophy. One of the earliest and most famous is the idea of the social contract, which was invented by British philosopher Thomas Hobbes in the seventeenth century. According to Hobbes, human beings originally lived in a "state of nature" in which life was "solitary, poor, nasty, brutish, and short." At that time, there was no human society, and individuals were in a perpetual state of war with one another. To alleviate these evils, according to Hobbes, they entered into a "social contract" in which they gave their liberty into the hands of a sovereign. Interestingly, Hobbes regarded the resulting society as a form of artificial life. Hobbes considered the mechanical automata which were constructed in his day to be examples of artificial life, too. One can't blame him. Some of those that were constructed in his day were quite impressive. Some time after Hobbes' death, the Swiss inventor Jacques de Vaucanson constructed an automaton which even some contemporary scientists might consider to be an example of artificial life. de Vaucanson's creation was a mechanical duck that quacked, flapped its wings, splashed about in water, drank, ate, digested its food, and defecated. Isn't there a saying that begins, "If it quacks like a duck . . . ?" As far as I know, Hobbes never used the term "artificial life." But the British philosopher Bertrand Russell did when writing about Hobbes in his 1945 book *A History of Western Philosophy*. Ideas—the concept of artificial life is only one example—sometimes turn out to be older than we think.

Modern anthropologists do not believe that Hobbes' "state of nature" ever existed. Human beings are social animals, after all. Even

Homo erectus formed societies. Hobbes probably didn't believe in a state of nature either. He used it as part of a thought experiment, which led to an argument intended to justify the absolute sovereignty of monarchs. Monarchy, he maintained, was necessary.

Few of us are royalists today. But what are we to make of the argument? One could attempt to refute it by creating another thought experiment in which some system other than a monarchy is formed. If one thought experiment leads to one conclusion and a second to a different conclusion, has anything really been gained?

On the other hand, suppose that we set up an artificial life experiment in which the agents represented individuals in a state of nature. One could then run the simulation and see what kinds of societies were created. There is no reason why this could not be done in the very near future. Anthropologists are already studying artificial societies.

Perhaps performing such an experiment would lead to no definite conclusions either. The critics would surely charge that the only knowledge gained was knowledge about what goes on inside a computer. But a certain amount of rigor would be preserved. No one could argue that the assumptions—either explicit or implicit—made at the beginning did not lead to the conclusions reached. As Dennett points out, "what 'stands to reason' or is 'obvious' in various complex scenarios is quite often an artifact of the bias and limitations of the philosopher's imagination more than the dictate of genuine logical insight." No one accuses computer programs of being biased, after all. According to the old cliché, if garbage is fed in, we get "garbage out." But that is the programmer's fault, not the computer's.

Dennett concludes that artificial life experiments are "a great way of confirming or discomfirming many . . . intuitions and hunches in philosophy. Philosophers who see this opportunity will want to leap into the field, at whatever level of abstraction suits their interests, and gird their conceptual loins with the simulational virtuosity of computers."

Furthermore, Dennett believes there are many questions "of manifest philosophical interest" that merge with the questions of philosophy. Some examples that he cites are:

- Why is there sex?
- Are there fixable scales or measures of complexity or desig-
nedness or adaptiveness that we can use to formulate hypotheses about evolutionary trends?
- Under what conditions does the fate of groups as opposed to
individuals play a decisive role in evolution?
- What is an individual?

One doesn't have to be a philosopher to be interested in such
questions. It seems to me they fall more into the realm of science
than that of philosophy. As a philosopher, Dennett is interested in
the philosophical aspects of things. Whether that is his intention or
not, he seems to be suggesting that research in artificial life is capable
of answering many of the important questions of biology.

A Strange Kind of Science

The sciences of complexity are an odd kind of science. Though I
bow to common usage and speak of the "sciences" of complexity, I
tend to think of it as a single field. To be sure, there exist sub-
disciplines, just as there are in physics, mathematics, astronomy, and
so on. But the field of complexity is not yet as fragmented as many
others. Nowadays, a condensed matter physicist and an astrophysi-
cist could not talk to one another about their respective work and
expect to be understood. This is not yet true in the sciences of complexity. Though different scientists pursue different kinds of research, they still speak a common language.

The traditional sciences, such as physics and chemistry, concern
themselves with natural phenomena and attempt to explain them.
Physicists study such things as the behavior of the components of
matter, superconductivity, transitions between liquid and solid
states, particles that are emitted in supernova explosions, and so on.
Chemists study such things as chemical reactions and bonds, and
molecular structure. Meteorologists study the weather patterns that
are seen on the surface of the earth. Ecologists study the interactions
between species in various kinds of environments. The scientists

who work in the sciences of complexity study only what they themselves have created.

The study of complexity is a scientific field that has no subject matter. To be sure, there are theories of complex systems, and attempts are made to create computer models of natural phenomena. In the end, the only things researched are phenomena that emerge in computers. Even if one is attempting to analyze something like traffic flow or fluctuations in financial markets, one looks at what is happening in the computer simulation, not at what is observed in the external world.

When I say this, no criticism should be inferred. The sciences of complexity may have spawned some strange science, but it is not as unique as many would be tempted to think. The devising of models of reality has always been an important part of scientific activity. For example, experimental physics is sometimes said to have begun with Galileo. Indeed, Galileo did place an emphasis on the experiment, an emphasis largely absent from the work of his predecessors, who were primarily concerned with explaining natural phenomena in terms of the teachings of Aristotle. As often as not, Galileo depended, not on experimental work that he had actually performed, but on thought experiments. He would describe an imaginary situation and declare that such and such a thing *must* be so. He probably never performed the legendary experiment in which two weights were dropped from the Leaning Tower of Pisa (which was described by a disciple of Galileo long after his master had died). But Galileo did invent imaginary situations involving falling objects and draw conclusions from them.

In some cases, Galileo seems to have thought that matters were so clear that no real experiment had to be done. In others, he was hampered by the lack of the requisite experimental technology. If he made use of models that he created in his mind, we do not blame him for it. On the contrary, we honor him as the first great physicist.

The physicists who developed quantum mechanics also made extensive use of thought experiments when they were trying to elucidate the fundamental principles of the new theory. As we know now, they did this with extraordinary success. Some of their "imaginary" experiments can now be performed. Almost without exception, the results are just as they predicted.

Science has always made use of models of reality. Traditionally, these have been models that were created and examined within the mind. During the last few decades, the development of high speed computers has made it possible to create more complex models and to examine them more deeply. When I say this, I am naturally not implying that the computers are somehow superior to the human mind. What I am saying is that the mind is able to make use of the calculating ability of computers to open up new, unexplored possibilities.

Are the thought experiments of Galileo and the evolution of artificial life-forms on the Internet really analogous to one another? I think that in a sense they are. They are both attempts to create artificial models of the natural world. They are also very different. Sometimes quantitative differences (e.g., speed of computation, complexity of the model being studied) translate into qualitative ones. In some respects, a human being is not very different from a bacterium. Both organisms make use of the same genetic code, and the enzymes in their cells cause RNA to create proteins in the same ways. Humans and other complex beings don't even have a cellular chemistry that is very much more complex than that of bacteria. Most of the enzymes that have been discovered in the cells of multi-cellular plants and animals are used by bacteria as well.

Nevertheless, quantitative differences exist. Humans have many more genes than bacteria do. This numerical difference is augmented by the fact that, if you have more genes, the number of possible gene networks is multiplied enormously. A number of mathematical theorems explain that when the number of objects in a set is—for example—doubled, the number of possible interactions between objects in the set increases by a factor that is much greater than two. It is this that allows human beings to have several hundred different cells types and numerous subtypes, and to possess such complex structures as the central nervous system.

We would expect that now that we no longer have to depend upon the mind alone, surprising new kinds of "thought experiments" should be created. Indeed, this is exactly what has happened. And there is no end in sight. Scientists are only just beginning to explore the possibilities inherent in the field of artificial life.

It has often been said there is a notable lack of experimental support for the ideas expressed by scientists working in the sciences of complexity. It has also been suggested that many different kinds of computer models would give similar results if "flexed in the right manner." Indeed, these are charges that have to be taken seriously. Until more contact is made between theory and experiment, we have no way of knowing what ideas are most likely to be true or which models are most likely to be accurate.

We should remember that time lags between the creation of new models and their empirical confirmation have been quite common in science. It was in the early seventeenth century that Galileo insisted the earth rotated. But it wasn't until 1851, more than two centuries later, that an experiment was performed in which it was conclusively demonstrated the earth rotated. In that year the French physicist Jean Bernard Léon Foucault showed that the apparent path of a heavy pendulum (known today as "Foucault's pendulum") swinging back and forth would shift as the earth rotated beneath it. Quantum mechanics was discovered in 1926 and developed during the years that immediately followed. Only today can some of the experiments that the early quantum physicists conceived actually be performed.

One should expect, then, that there should be some lag between the introduction of new ideas in the sciences of complexity and their confirmation. It is somewhat disappointing that there are few so far. One sometimes has the impression that scientists in the field are much more interested in devising new computer simulations than they are in finding out whether or not they have any relevance to the real world. There is every reason to think this situation will eventually be remedied.

In the meantime, new and original ideas seem to arise within the field almost every day. Anyone who writes about the sciences of complexity cannot help but be infected with some of the feeling of excitement that has been generated. This sense of excitement is only heightened by the feeling that it is the sciences of complexity which will finally give us an answer to that venerable question, "What is life?"

BIBLIOGRAPHY

I have tried to select books accessible to the general reader. There are a few exceptions, which are noted.

Adami, Christoph. 1998. *Introduction to Artificial Life*. New York: Springer–Verlag. This book is definitely too technical for the average reader (and for most biologists as well). It is a textbook that requires a background in statistical physics, thermodynamics, and basic biology. I thought I should list it because much of the material deals with Adami's artificial life program, Avida, which is discussed in the text. A CD-ROM accompanies Adami's book and contains Avida software which can be run on Windows 95, Windows NT, and Unix operating systems (but not on the Macintosh). Avida can also be downloaded from Adami's web site (see the "World Wide Web Resources" section).

Brockman, John. 1995. *The Third Culture*. New York: Simon & Schuster. Literary agent John Brockman taped interviews with a number of leading scientists, many of whom have worked in evolutionary biology or in the sciences of complexity. He edited the interviews to make them read like essays. Also included are comments by these scientists on one another's work. Contributors include Stephen Jay Gould, Richard Dawkins, Brian Goodwin, Niles Eldredge, Daniel Dennett, Stuart Kauffman, Chris Langton, and Doyne Farmer.

Bury, J. B. 1932. *The Idea of Progress.* **New York: Macmillan, 1932.**
From the very beginning, evolution was erroneously equated
with the idea of progress toward "highly evolved" beings. Perhaps it was only natural that this should have been the case, for
the idea of progress was almost a religion to the Victorians. This
classic book explains how the idea of progress evolved.

Darwin, Charles. 1859. *On the Origin of Species.* **London: John Murray.**
The original book; there are many recent editions.

Dawkins, Richard. 1976. *The Selfish Gene.* **Oxford: Oxford University Press.** Dawkins expounds the idea that evolution is a result of genes that "want" to leave as many copies of themselves
in succeeding generations as possible. Naturally he doesn't really believe genes have desires; he is only using a colorful metaphor. Dawkins' point of view is orthodox Darwinism; he doesn't
dispute the fact that natural selection acts on individual organisms, not on genes.

Dawkins, Richard. 1986. *The Blind Watchmaker.* **Harlow, U.K.:
Longman.** Dawkins discusses the role of chance in evolution,
and reiterates his ideas about "selfish genes."

Dawkins, Richard. 1996. *Climbing Mount Improbable.* **New York:
Norton.** The title might lead one to think this is a book about
fitness landscapes. It isn't, though a few brief references are
made to them. Dawkins is more concerned about the evolution
of seemingly improbable biological structures and behavior.

Dennett, Daniel C. 1995. *Darwin's Dangerous Idea.* **New York: Simon & Schuster.** This is an exposition of the "ultra-Darwinist"
point of view, followed by speculations on philosophical topics.
Dennett argues that those who will not accept "Darwin's dangerous idea" of natural selection are afraid of it because it seems
to threaten humanistic values. Even those who react adversely
to Dennett's *ad hominem* style of argument will find a great deal

to ponder in this book. Naturally there is a lot of commentary of the work of Dennett's *bête noir*, Stephen Jay Gould.

Dennett, Daniel C. 1998. *Brainchildren*. Cambridge, MA: The MIT Press. This book contains Dennett's essay, "Artificial Life as Philosophy," which first appeared in the journal *Artificial Life*. The bulk of the book consists of essays by Dennett on his main interest, artificial intelligence. If you're interested in the essay, and are connected to the Internet, it might be simplest to download it (see the "World Wide Web" resources section that follows this bibliography). It's not very long. You can't go wrong getting the book either. Dennett is much more readable than most contemporary philosophers. I think this is because he likes to invent a lot of thought experiments (including ones about brains in vats), and doesn't engage in closely reasoned arguments (which may or may not be correct) as often as other philosophers do.

Eldredge, Niles. 1995. *Reinventing Darwin*. New York: John Wiley. Eldredge discusses his and Stephen Jay Gould's ideas about punctuated equilibrium, and the scientific debates their theory has inspired. There is a lot of emphasis on the idea that there is more to evolution than the action of natural selection on individual organisms. According to Eldredge, evolution also occurs at high levels of complexity.

Frank–Kamenetski, Maxim D. 1993. *Unraveling DNA*. New York: VCH Publishers. If you're interested in knowing something about molecular biology, you generally have two choices: read a textbook, or read a book which deals more with the discovery of DNA than its structure or functions. This book, originally published in Russian in 1988, is an excellent choice for the nonscientist. It's a sort of introductory-level textbook, but it is always interesting and not difficult to read (now if someone would only do the same for cell biology!).

Gell–Mann, Murray. 1994. *The Quark and the Jaguar*. New York: Freeman. Nobel laureate Gell–Mann discusses the two fields that have interested him the most: high-energy particle physics and the sciences of complexity. Gell–Mann writes the deepest discussion of the meaning of the term "complexity" that I have run across. Among other things, he relates mathematical definitions of the term to complexity in the real world, and finds that none of them are really quite adequate. This isn't very surprising. Hundreds of definitions and measures of complexity have been proposed. However, Gell–Mann's discussions give the reader real insights into the difficulty of the problem.

Goodwin, Brian. 1994. *How the Leopard Changed Its Spots*. New York: Scribners.   Goodwin expounds his ideas about biological forms that are not the result of natural selection. The subtitle, "The Evolution of Complexity," is a reasonably good description of this book.

Gould, Stephen Jay. 1989. *Wonderful Life*. New York: Norton.   Gould discusses the Cambrian explosion and the discovery, interpretation, and reinterpretation of the fossils that led to the realization that something unique took place 530 million years ago. Although Gould's writing sometimes gets a little technical, his discussions are always interesting and relatively easy to follow.

Gould, Stephen Jay. 1995. *Dinosaur in a Haystack*. New York: Harmony Books.   Gould, who writes a column for *Natural History* magazine, periodically collects his essays into a book. Each volume provides an excellent education in such subjects as paleontology and evolutionary biology. At present, this is the latest installment. Other books in the series include *Ever Since Darwin* (Norton, 1977), *The Panda's Thumb* (Norton, 1980), *Hen's Teeth and Horses' Toes* (Norton, 1983), *The Flamingo's Smile* (Norton, 1985), *An Urchin in the Storm* (Norton, 1988), and *Bully for Brontosaurus* (Norton, 1992).

Gould, Stephen Jay. 1996. *Full House*. New York: Random House.   Gould expounds the idea that there is no trend toward increasing complexity in evolution. There is, somewhat surprisingly, a lot in here about baseball. Gould uses it to explain certain mathematical ideas.

Kauffman, Stuart. 1993. *The Origins of Order*. Oxford: Oxford University Press. This has been called "a monster of a book." Even scientists sometimes have trouble following all that Kauffman says, if they lack the proper mathematical background. Nevertheless, it is the most complete exposition of Kauffman's theories.

Kauffman, Stuart. 1995. *At Home in the Universe*. Oxford: Oxford University Press. A semipopular treatment of the ideas expressed in *The Origins of Order* (see previous reference).

Langton, Chris, ed. 1997. *Artificial Life: An Overview*. Cambridge, MA: The MIT Press. The first three issues of the journal *Artificial Life* are collected in this volume. Most of the contributions are rather theoretical. The emphasis is on papers defining artificial life as a field rather than on research results. Langton has also edited the proceedings of a number of artificial life conferences. The first three volumes were published by Addison–Wesley, the proceedings of the fourth and fifth by MIT Press. Langton also edits the journal *Artificial Life*, which is published by MIT Press. The material in these publications can get pretty technical.

Levy, Steven. 1992. *Artificial Life*. New York: Random House. Levy discusses the development of the field of artificial life. He covers relevant topics, such as cellular automata. This is another book that is very easy to read. It provides an excellent overview of the earlier work in the field.

Lewin, Roger. 1992. *Complexity*. New York: Macmillan. Lewin discusses the sciences of complexity and the people who did the most to develop the field. He tells many anecdotes about these scientists' personal lives. I found especially interesting the story of a hang gliding accident experienced by Chris Langton. Langton suffered brain damage and other injuries as well. As he was

recovering, he found that parts of his personality would suddenly return bit by bit. At the very beginning of his recovery, he
knew that something was missing, but he didn't know exactly
what it was. He eventually regained all his faculties.

Malkiel, Burton G. 1976. *A Random Walk Down Wall Street*. **New
York: Norton.**   In this now classic book, Malkiel argues that
stock market fluctuations are essentially random, that no one
can predict the future price movements of any security. The
author maintains that a broad portfolio of stocks chosen at random will do as well as one carefully chosen by experts. Malkiel
analyzes only stock markets, but his idea would apply to other
kinds of financial instruments as well. Mathematicians and scientists tended to adhere to this view until analyses by Doyne
Farmer and Norman Packard, and the success of the Prediction
Company, showed there were some chinks in Malkiel's theory.

Murphy, Michael P. and O'Neill, Luke A. J., eds. 1995. *What Is Life?
The Next Fifty Years*. **Cambridge: Cambridge University Press.**
None of the contributors (who include Nobel laureate Manfred
Eigen, Stephen Jay Gould, Stuart Kauffman, and John Maynard
Smith) really tries to answer the question, "What is Life?" How-
ever, their thoughts are quite interesting. The subtitle, "Specula-
tions on the Future of Biology," is a pretty good description of the
book.

Pagels, Heinz R. 1988. *The Dreams of Reason*. **New York: Simon &
Schuster.**   This may have been the first book on complexity,
and it is still one of the best. There is more here about artificial
intelligence than there is about artificial life. This isn't surprising, for the former was the more highly developed field at the
time.

Silk, Joseph. 1997. *A Short History of the Universe*. **New York:
Scientific American Library.**   The universe has evolved, too. A
few hundred thousand years after the big bang, it consisted of
little but clouds of hydrogen and helium gas. Today, there are

stars, galaxies, planets, and life. I wouldn't suggest that cosmic evolution is perfectly analogous to biological evolution, however. For one thing, the universe did not evolve by natural selection. However, I suspect that anyone interested in biological evolution will be interested in finding out how the other kind works. Silk is a professor of astronomy and physics at the University of California at Berkeley. His book is just one of many excellent ones on cosmology currently available. I selected Silk's book because he is a knowledgeable author, and because it just happens to be the last book on the subject that I have read, and a very good one at that.

Wesson, Robert. 1991. *Beyond Natural Selection*. Cambridge, MA: The MIT Press. In this extensively researched book, Wesson makes a case for the contention that natural selection cannot be all there is to evolution. He says little about complexity, however. He is more concerned with amassing evidence that supports his point of view.

WORLD WIDE WEB RESOURCES

The preface "http://" is left off of the following web addresses. Most web browsers do not require that you type it in. Thus the full address of a site given as www.santafe.edu is actually http://www.santafe.edu. The periods that appear at the end of web addresses are there for punctuational purposes; they are not a part of the addresses themselves.

There are many sites with Alife games listed here. Most of them are just that: games. Though created with serious purposes in mind, they have nothing near the complexity of such programs as Tierra and Swarm, or of commercial games, for that matter. If you peruse the sites, you will find a number of Java applets. Java applets are animations meant to be viewed through your World Wide Web browser. They take the name from the programming language they are written in called Java.

Please note whenever I say software can be downloaded for "free," I'm talking about "freeware" as opposed to "shareware." You are not expected to pay for freeware if, after downloading and trying it, you decide to keep it on your computer. Nowadays many shareware programs cease to function after some period of time (usually thirty days) or a certain number of uses. This is not the case with freeware programs; they're yours to keep.

Finally, you'll find a lot of whimsy here. Artificial life researchers find nothing wrong with mixing a little humor in with their science.

Alife Page **alife.fusebox.com.** This is an artificial life games site. You can watch or play such games as "Flames," "Psychedelic Demons," "Hungry Swarm," and "Planet Wator." You can also enter the Morphs Lab and answer a somewhat frivolous quiz on the "Meaning of Life."

Amazon.com **www.amazon.com.** This is the online bookstore. It is a good place to search for books on complexity or any other subject that interests you. I was impressed with the breadth of their offerings. I discovered, for example, that they listed all of my out of print science books, as well as those that were in print, and a couple of collections of my poetry as well. You're likely to find some items you, too, didn't expect to find. Their largest competitor is Barnes & Noble (www.barnesandnoble.com), which does a good job too.

Applets for Neural Networks and Artificial Life **www.aist.go.jp/ NIBH/~b0616/Lab/Links.html.** A lot of the material that you'll find here is intended for scientists, but don't let the name intimidate you. This site has, among other things, a compilation of Alife simulations and Alife games you can play online. You'll find a link to Boids here. There is also a blackjack game in which you are given the opportunity to compete against a computer program which is capable of learning and refining its play. Because the site is a collection of links, I have listed a number of them separately. That way, you can go directly to the blackjack game, for example, without having to look for the link. But you'll look for "Test of Unicursality of a Graph (in Japanese)" in vain; when I was compiling the separate listings, I figured that I could probably skip that one.

Artificial Life Games Homepage **gracco.irmkant.rm.cnr.it/luigi/ lupa_algames.html.** Downloadable artificial life games.

Avida Artificial Life Group **www.krl.caltech.edu/avida/.** Lots of information on the Avida artificial life project. There is also downloadable software.

Beasties   **animas.frontier.net/~srladd/beasties.html.**   Perform evolutionary experiments online.

Biotopia   **alife.santafe.edu/~liekens/biotopia.html.**   Software that creates artificial Darwinistic ecosystems.

Biological Model Simulation Project   **www.sfc.keio.ac.jp/~t94809ms/ project.html.**   You'll find some programs here which simulate the behavior of schools of fish in the presence of a predator and when they're being fed. The comments on the programs are written by some Japanese scientists whose command of English is less than perfect. But don't despair; there's also a Japanese version, which is presumably more grammatical.

Boids   **hmt.com/cwr/boids.html.**   A site devoted to Craig Reynolds' Boids. There is even a Boids screen saver available here.

Bumble   **animas.frontier.net/~srladd/bumble.html.**   Artificial bees that evolve.

Center for Cognitive Studies   **www.tufts.edu/as/cogstud/papers/ mainpg.html.**   You can find Daniel Dennett's essay on "Artificial Life as Philosophy" here. Numerous other papers by Dennett can be downloaded. However, he is more interested in artificial intelligence than in artificial life.

Complexity On-Line   **www.csu.edu.au/complex.**   A scientific information network about complex systems.

Evolution of Critters   **www.alumni.caltech.edu/~croft/java/Evolve/ Evolve.html.**   As the name implies, this is an evolution simulation program. It's a rather simple one, but interesting.

Evolutionary Computation and Artificial Life   **www.ai.mit.edu/ people/unamay/EC.html.**   There are lots of links here. Much of the material you'll find in them is rather technical.

Fishtank **www.morganmedia.com/m2/shock.html.** A real time animated fish tank for Netscape 2.0. Though I haven't tried it, I imagine that this one will work just as well in newer versions of Netscape.

Floys **www.aridolan.com/JavaFloys.html.** Floys are flocking (Boids-type) artificial animals that defend their territory against intruders. If you find Floys interesting, you might also want to connect to the iFloys page at www.aridolan.com/iFloys.html. The difference between Floys and iFloys is that the former all have the same rules of behavior, while the latter can have individual traits. This makes it possible to perform some experiments in social behavior (yes, you can do this online). For example, if one iFloy is given nervous or energetic traits (high speed and acceleration), the whole group becomes a little confused and disorganized.

Gecko **peaplant.biology.yale.edu:8001/papers/swarmgecko/rewrite.html.** Gecko is the Swarm-based successor to Echo. It is used for ecological simulations.

Stephen Jay Gould. Gould does not use a computer (he may have the right idea; at least he doesn't have to spend time on technical assistance lines getting his typewriter to work), so naturally he doesn't have a website. But a web search will turn up numerous references to Gould and his work.

Helix **www.necrobones.com/alife.** Helix is a Tierra-like system for Windows.

Stuart Kauffman **www.santafe.edu/People/kauffman.** Kauffman's home page.

Macintosh Artificial Life Software **www.ccnet.com/~bhill/else-where.html.** Free downloads.

Manna Mouse **www.caplet.com/MannaMouse.html.** This site contains an interactive game that requires you to construct a fitness landscape. Some Alife games are fun to watch or play. This one is more like a homework exercise. Still, if you're feeling really dedicated. . .

Martian Meteorite **cass.jsc.nasa.gov/lpi/meteorites/mars_ meteorite.html.** One of the reasons for creating artificial life is the fact that terrestrial life is the only kind that we previously had to study. There is, of course, a third possibility: extraterrestrial life. If you're interested in the claim that a meteorite found on Earth contains traces of life that once existed on Mars, take a look at this NASA site.

Prediction Company **www.predict.com.** Doyne Farmer's and Norman Packard's investment company.

Primordial Soup Kitchen **psoup.math.wisc.edu/kitchen.html.** In addition to lots of stuff on artificial life, you'll find some good recipes at this site (but I still like my recipe for split pea soup better).

Rat Maze **www.sce.carleton.ca/netmanage/java/Maze.html.** Ever feel like you're a rat running a maze?

Red Hat **www.redhat.com.** There are a lot of places where you can download Linux for free. Red Hat expects you to pay for its version. However, they make the installation process relatively easy and include a number of useful utilities.

Santa Fe Institute **www.santafe.edu.** There are a lot of resources here, some of which are individually listed elsewhere in this section. Artificial life software, information about current research, links to the web pages of individual scientists are among the items you will find. Many of the papers here are available for download only in PostScript format. If you don't have PostScript, you can view and print them with Ghostview and Ghostscript, respectively. Both are available free at numerous different sites on the Web. You'll generally find them bundled together.

Scripps Research Institute **www.scripps.edu.** The Scripps Institute is a major center for biological research. Reza Ghadiri's and Julius Rebek's web sites can be found here. Ghadiri's contains a number of graphics that depict the molecules he works with.

6th Alife Conference **www.cs.ucsd.edu/~rik/alife6/index.html.** The sixth Artificial Life Conference was held at UCLA in June 1997. This is the conference website. Included among the materials available here are abstracts (one-paragraph synopses) of the papers given at the conference. The material is somewhat technical. Nevertheless, perusing the site is a good way of getting an idea of the kinds of research currently being done in the field. Links to the home pages of the scientists who gave papers at the conference are often provided.

Swarm **www.santafe.edu/projects/swarm.** The Swarm website. There is a lot of material here, including descriptions of Swarm, information about the people associated with the project and their research, downloadable software, and information on the research projects of people at other educational institutions who are using Swarm.

Technology Review **www.mit.edu/afs/athena/org/t/techreview/ www.** You can find Robert Crawford's article, "What's It All About, Alife" here.

Tierra **www.hip.atr.co.jp/~ray/tierra/tierra.html.** The Tierra web site also contains a great deal of information, including ongoing progress reports on the Tierra network project.

World of Richard Dawkins, The **www.spacelab.net/~catalj.** Dawkins is not associated with this site. There is an enormous amount of interesting material on Dawkins and his books here.

Zooland **alife.santafe.edu/~joke/zooland.** A collection of artificial life resources maintained by the Santa Fe Institute. And why
does the word "joke" appear in the address? The creators of
this site say, "And don't take anything too seriously; it's only
science."

INDEX